Reliability and Ecological Aspects of Photovoltaic Modules

Edited by Abdülkerim Gok

Published in London, United Kingdom

IntechOpen

Supporting open minds since 2005

Reliability and Ecological Aspects of Photovoltaic Modules
http://dx.doi.org/10.5772/intechopen.82613
Edited by Abdülkerim Gok

Contributors
Titu-Marius I. Băjenescu, Rüştü Eke, Mahmoud Zendehdel, Narges Yaghoobi Nia, Mohammadreza Yaghoubinia, Hervé Joël Tchognia Nkuissi, Fransisco Konan, Bouchaib Hartiti, Jean-Marie Ndjaka, Leonimer Melo, Dario Toginho, Alex Archela, Mohamad Kharseh, Holger Wallbaum, Laurentiu Fara, Dan Craciunescu, Claudia Barolo, Nicole Mariotti, Matteo Bonomo

First published in London, United Kingdom, 2020 by IntechOpen
IntechOpen is the global imprint of INTECHOPEN LIMITED, registered in England and Wales, registration number: 11086078, 7th floor, 10 Lower Thames Street, London, EC3R 6AF, United Kingdom
Printed in Croatia

British Library Cataloguing-in-Publication Data
A catalogue record for this book is available from the British Library

Additional hard and PDF copies can be obtained from orders@intechopen.com

Reliability and Ecological Aspects of Photovoltaic Modules
Edited by Abdülkerim Gok
p. cm.
Print ISBN 978-1-78984-822-9
Online ISBN 978-1-78984-823-6
eBook (PDF) ISBN 978-1-83968-452-4

An electronic version of this book is freely available, thanks to the support of libraries working with Knowledge Unlatched. KU is a collaborative initiative designed to make high quality books Open Access for the public good. More information about the initiative and links to the Open Access version can be found at www.knowledgeunlatched.org

Meet the editor

Dr. Abdülkerim Gok is a research associate in the Department of Materials Science and Engineering at Gebze Technical University, Turkey. After completing his BS in Materials Science and Engineering at Anadolu University, Turkey in 2007, he received his MSc in Chemical Engineering from Columbia University, New York, USA, in 2011, and his PhD in Materials Science and Engineering from Case Western Reserve University, Cleveland, Ohio, USA, in 2016. His research interests include performance, lifetime, and degradation science of photovoltaic (PV) modules, the effect of environmental stressors on PV module materials, predictive and diagnostic degradation pathway models, and reproducible statistical methods.

Contents

Preface

In order to meet growing energy needs and compete effectively with conventional energy resources and hence reduce greenhouse gas emissions, photovoltaic (PV) solar energy is considered a form of renewable, sustainable, and clean energy resource. Due to rapid development and reduced costs, the last few years have seen tremendous increase in PV module installations. With new policies, financial incentives, and competitive tenders, global installations are expected to continue growing. PV modules consist of many parts and any failure during operation puts the integrity and functionality of modules at risk. Not only does failure jeopardize power production, but it also impairs insulation and induces safety hazards. Developing new solar cell materials and module components, and understanding and preventing degradation and failure mechanisms in current technologies, are essential for improved efficiency, performance, and service lifetime of PV modules. There are also growing concerns over the environmental impact of PV modules due to the toxic chemicals used or created during manufacturing processes, solar waste discarded at the end of life, and recycling problems. While trying to compete with fuel-based traditional energy resources to reduce ecological pollution and global warming emissions, the potential environmental impacts associated with PV modules should be mitigated by taking sustainability measures and raising public awareness. This book reviews some of the failure mechanisms and reliability issues observed during indoor testing and outdoor service and then provides useful insights into ecological concerns for greener technology.

Abdülkerim Gok
Gebze Technical University,
Turkey

Section 1

Field Performance

Chapter 1

PV Outdoor Tests

Rustu Eke

Abstract

The main purpose of this chapter is to survey the structure, operation, and design of photovoltaic (PV) systems. PV systems consist of solar cells and electronic units which convert directly produced electricity from solar irradiation to electricity in the form of demand by load or feed the produced electricity directly into the grid. The heart of the system is the solar cell or PV array. From individual solar cell to PV power plant and solar electricity conversion will be discussed in this chapter. Indoor and outdoor measurement of PV modules and performance of PV systems will be summarized. The performance of the system which is mainly the energy output depends on the operating condition, the location of the system, and the configuration of the system. The system modeling and its behavior under varying weather conditions which strongly affect the electricity output of the system will be discussed in this chapter.

Keywords: PV module, PV system, performance, indoor and outdoor measurements, PV electricity cost

1. Introduction

During the last few decades, there has been an exponential growth in photovoltaics across the world. Although market grows day by day and correspondingly employment rate increases, this brings many problems associated with the quality of the system due to several factors [1]. PV system installations have increased, and now in annual installations, PV is one of the leading power capacity additions. In 2018, over 100 GW of new PV power capacity was added. The annual PV capacity addition in 2018 was more than the total installed capacity in 2012. Total installed PV power capacity was in excess of 500 GW at the end of 2018 [2]. The power produced by a PV system depends on a range of factors which need to be examined when the system is designed [3]. These factors can be given such as operating conditions, the details of the configuration of the system, the location of the system, the amount of received solar radiation, the ambient temperature, and other climate-related aspects.

This chapter provides an introduction to the PV system configuration and the influences of these parameters on PV system performance.

2. PV system

There are two main classifications of PV systems. The first one is grid-connected where PV modules produce their maximum energy and they always feed the produced electricity in the form of local electricity grid. The other is stand-alone PV systems that operate independent from the grid, and they supply the electricity for the specified load. In these types of systems, PV modules do not operate at their maximum. Thus, power and generated energy values are limited with the capacity of storage. Grid-connected PV system schematic is simple and given in **Figure 1**.

The grid-connected system is often classified into two as distributed and centralized systems. Small systems are generally distributed and have a capacity less than 100 kW. Most of these systems are installed on roofs or at the top (garage, patio, winter garden, etc.) or beneath the buildings. Although distributed systems are connected to low-voltage grid and meet the local load centralized PV systems which are connected at a higher voltage, the main purpose of them is feeding the general grid supply. There has been an increase on the side of grid-connected ratio since 2009 because of the high-efficiency ratio of the PV system with respect to stand-alone PV systems as well as simplifications and improvements in grid connections. According to the IEA PVPS data, grid-connected PV systems represented around 62% of the cumulative installed PV capacity at the end of 2017 where this ratio is only 22% in 2009 [4].

The stand-alone system operates independently from the grid and provides the power and electricity of the specified load or loads. There is a charge controller and a battery bank different from the grid connected to the PV system. Moreover the inverter operates in a different way. The charge controller controls the charging and discharging of batteries and consists of a maximum power point (MPP) tracker for operating PV modules at a maximum power. A schema of stand-alone PV system is given in **Figure 2** with different loads.

The hearth of the PV system is the solar cell itself where a range of semiconductors are used in solar cells. PV modules have to offer a high performance, a stability in operation, and good and low-cost manufacturability, and they have to perform a long lifetime. Electricity yield is important for PV modules. Currently the installations in the established PV market are dominated by crystalline silicon (c-Si, including mono- and multicrystalline silicon). Other commercial PV technologies in the market are cadmium telluride (CdTe), copper indium gallium diselenide (CIGS), amorphous silicon (a-Si), and several hybrid designs. There are also different types of solar cells like organic and polymer-based cells and some multi-junction cells. Some of them are in the market with a small ratio, and they are classified in emerging PV technologies [5]. In terms of performance, the most important difference between the module types is the conversion efficiency which is the ratio of the electrical output and the amount of solar irradiation received to the solar cell or module plane. There is a continuing development in cell and module efficiencies under tests, and the results are periodically publicized in some journals, and the latest solar cell and module efficiencies are summarized in **Table 1** [6].

The other parameters affecting the PV system performance are spectral distribution of light, temperature coefficients of PV module, cell stability, encapsulation quality, shading effect, design of PV modules (wiring of PV modules, number of series-parallel-connected PV modules), and other components like inverter and wiring other than the PV modules, namely, balance of system (BOS) equipment.

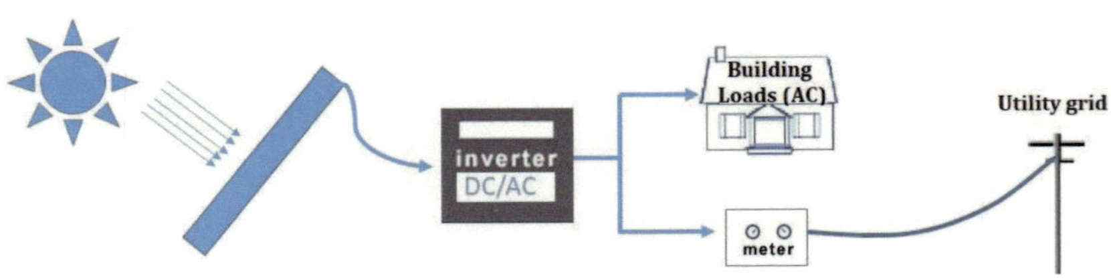

Figure 1.
Schematic diagram of a grid-connected PV system.

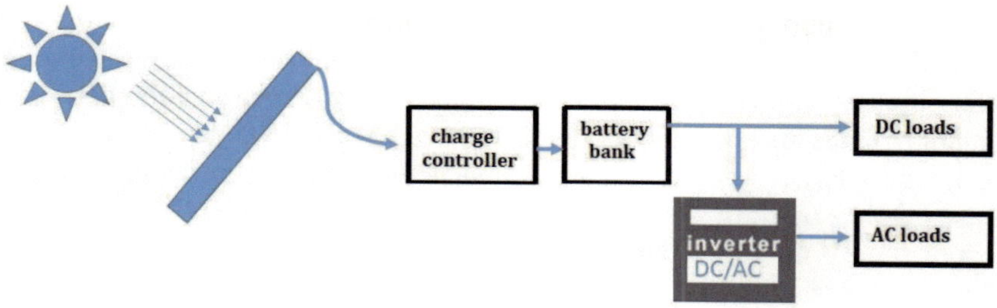

Figure 2.
Schematic diagram of a stand-alone PV system.

Module technology	Efficiency range under standard test conditions, large area (module efficiency) (%)	Highest reported laboratory efficiency, small area (%)
Crystalline silicon (sc-Si)	16–24.4	26.7
Multicrystalline silicon (mc-Si)	15–19.9	22.3
Amorphous silicon (a-Si)	6–11.9	14.0
CdTe	14–18.6	21.0
CIGS	15–19.2	22.9
Multi-junction	21–31.2	32.6
Organic	6–8.7	11.2

Table 1.
Solar cell and module efficiency ranges reported in late 2018.

3. PV system performance

Solar cells are the hearth of electrical conversion, and their electrical characteristics are similar with diodes. Therefore their current, I, and the voltage, V, relation will be given in Eq. (1):

$$I = I_o\left[exp\left(\frac{qV}{n k_B T}\right) - 1\right] - I_L \tag{1}$$

Here, I_o is the reverse saturation current of the diode, q is the electron charge $(1.602 \times 10^{-19} C)$, k_B I the Boltzmann's constant $(1.38 \times 10^{-23} \mathrm{m}^2\mathrm{kg}/s^2K)$, T is the operating temperature in Kelvin, I_L is the light-generated current, and n is the diode quality factor depending on cell material.

In real devices, some parasitic resistances have to be defined. R_s is the series resistance for representing resistances related to carrier transport and ohmic contacts in material. R_{sh} is the shunt or parallel resistance representing leakages.

Under these acceptances I-V characteristics of a real solar cell can be given by Eq. (2):

$$I = I_L - I_o\left[exp\left(\frac{q(V + IR_s)}{n k_B T}\right) - 1\right] - \frac{V + IR_s}{R_{sh}} \tag{2}$$

Solar cells can be connected in series or parallel to achieve higher current and voltage values. Only the values of current and voltage values can be changed in larger PV arrays. This equation can be applied to a variety of solar cell types. Only some parameters (like diode quality factor) will show different values which cannot be physically described easily. But mathematically the equation overlaps the experimentally obtained values. As power is the product of current and voltage ($P = I \times V$), the current-voltage and power-voltage relation of a typical PV device can be given in **Figure 3**, where P_{max}, I_{mpp}, and V_{mpp} values are power, current, and voltage values of the desired device operating at its maximum power point (MPP).

Most of the PV module performance parameters are given in Eq. (2), and the resulting characteristics are given in **Figure 3**. I_{sc} is the short-circuit current when there is no voltage across the device and it is nearly equal to I_L. V_{oc} is the open-circuit voltage when there is no flow of current and it is the maximum voltage available from the device. In order to obtain the maximum power from a device, it'd operate round maximum power point (MPP). One of the other performance parameters of a PV device is the fill factor (FF) which is the ratio of power at MPP to I_{sc} and V_{oc} product. MPP also defines the efficiency of the device, and the efficiency is the ratio of power output to incident power falling on the surface of the device (Eq. (3)):

$$\eta = \frac{Power\ output}{Power\ input} = \frac{P_{max}(FF I_{sc} V_{oc})}{irradiation \times A} \tag{3}$$

where A is the total area of the device [6, 7].

Under ideal circumstances, PV devices operate at MPP, but in real operating conditions, PV devices operate round MPP, and most PV systems have maximum power point tracking units to operate with a minimum loss of power available from the device. In electrical connection of PV modules in forming PV arrays, it is very important to connect identical PV modules in the same lines for the best performance. Sometimes while designing the PV array, there should be some restrictions in obtaining uniform irradiation on the same array. If the PV modules do not have uniform irradiation, the electrically series-connected units obey the weakest one, and this results in a decrease in the output and loss of performance.

The output of the PV devices changes under operating conditions, so PV modules are produced and launched to the market according to the power values of the device obtained at a standard set of operating conditions. These conditions are standard test conditions (STC): irradiance of 1000W/m^2, standard global spectrum at air mass 1.5 and operating temperature of 25°C [6]. These conditions are the only test conditions and rarely found outside in the operating conditions. Irradiance is the level that defines the energy input of the system, and it varies throughout the day and the season. Spectrum of the light defines the amount of generated current of the device, and there will be a difference in the spectrum although the irradiance will be the same. The operating

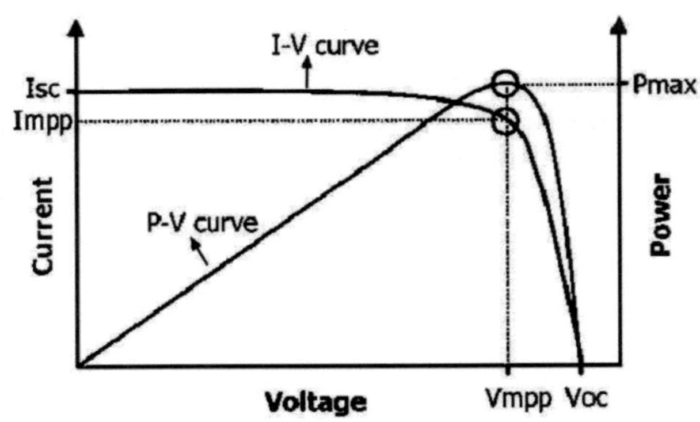

Figure 3.
The current-voltage and power-voltage characteristics for a solar cell/PV module/array.

temperature of the device is very variable, and it is generally over the value at STC. This high temperature depends on several factors like ambient temperature and plane of array solar irradiance level (G_{POA}) in W/m^2. Usually, the nominal operating cell temperature (NOCT) is given because of the difference at operating temperature in the field from STC values. These conditions are 800W/m² irradiance level, 20°C ambient temperature, 1 m/s wind speed, and under no load. NOCT values are usually given in PV module datasheets, and temperature is generally taken between 45°C and 50°C. The relation to find module operating temperature under NOCT is given in Eq. (4) [8]:

$$T_{module} = T_{ambient} + \frac{G_{POA}}{800}(NOCT - 20) \quad (4)$$

3.1 PV system efficiency and yield

PV device/module efficiency and performance parameters are given above. But a PV system consists of other components. For this reason, these components' efficiency values have to be taken into account while calculating the system efficiency. The end user is interested in the electricity produced by the PV system and its lifetime under real operating conditions because this directly influences the payback period or return time of the investment. The commonly used energy-rating standard for PV modules is given by the International Electrotechnical Commission (IEC) with IEC61853 (PV performance testing and energy rating part 1) [7]. Also the location of the system has to be considered while sizing the PV system. Energy yield measurements of PV systems at different climatic locations play an important role in the validation of the energy rating standard, deeper understanding of PV performance, and lifetime. Long-term and accurate measurements under real operating conditions are necessary, but there is currently no standard on how these measurements should be performed [9]. Most analyzers use inverter output or meter data for calculating the performance of PV systems. If the system is stand-alone, some other parameters like night loads and loss of load probability have to be regarded while designing the system. Energy yield and performance ratio (PR) values are the other parameters which defines the system's overall performance. Only PV efficiency is given in Eq. (3), but the power output of the whole system and calculation of efficiency using this output is very important, considering the other losses like wiring losses and inverter losses. In grid-connected PV systems, electricity output, solar resource, and system losses are the main parameters. Accurate evaluation of PV system performance is critical for PV industry.

Performance ratio defines the performance of the system [10, 11]. It is dimensionless and given as the ratio of $\boldsymbol{Y_F}$, final yield, and $\boldsymbol{Y_R}$ reference yield. In general PR value of the system is calculated in a monthly or yearly basis. Sometimes this value will be calculated for small intervals such as daily or weekly, but greater interval is preferred. As in some seasons because of the high module operating temperature and high irradiance values, PR values will be calculated high, and generally this value is between 0.60 and 0.80:

$$PR = \frac{Y_F}{Y_R}(dimensionless) \quad (5)$$

where the final PV system yield $\boldsymbol{Y_F}$ is the ratio of net electricity output, $\boldsymbol{E_{out}}$, of the PV system to the nameplate DC-installed power, $\boldsymbol{P_o}$

$$Y_F = \frac{E_{out}}{P_o}(\mathrm{kWh/kW})\,or\ (\mathrm{h}) \quad (6)$$

And the reference PV system yield Y_R is the ratio of H_t total plane of irradiance to G reference irradiance($\mathbf{e.g.\ 1000W/m^2}$):

$$Y_r = \frac{H}{G}(\text{h}) \tag{7}$$

Long-term PR calculation takes the system failures into consideration so it gives better results. But PR is neglected by some researchers, and only the electricity output per installed power is taken into account. Namely, they calculate only kWh/kWp ratio where kWp is the installed DC power capacity of the system [12].

Instead of defining the overall performance of the system, sometimes it is useful to consider the specific performance of a certain part of the system to find the correct design. Sometimes the output will not match the expectations. It will be PV array output or PV system output. In that situation, some modules, wirings, or other components will be analyzed. Most inverters give DC and AC power values at a certain time, so it is easy to calculate their exact efficiency. PV system operates during the daytime, so inverters also operate and their output values also vary. In semi-cloudy days, sometimes there should be a sudden decrease and increase in the irradiation level, and this causes a big difference in electrical value. Also the inverter efficiency varies sharply. At low irradiation levels, inverter efficiencies are low, but generally after 15% of their nameplate power, their efficiency is round 90%. Because of the variation in the irradiation level, input and output power, a new efficiency classification is defined in inverter efficiency calculation. It is Euro Efficiency or California Energy Commission (CEC) efficiency. Both are weighed efficiency values, and they use different efficiency values at different power input values and give lower than the peak efficiency but more representative values.

4. PV module characterization

The output of a PV system depends on various parameters, but one of the most important parts is the PV modules used. The electrical performance and output are given in Section 3 with module characteristics. Whether these characteristics are taken in laboratory conditions or in outdoor conditions, it is possible to translate the parameters to STC values and compare the nameplate values. A typical setup for laboratory I-V curve measurements is shown in **Figure 4**. The setup consists of a light source, a reference device for the determination of irradiance during the measurement, some temperature sensors, and an electronic load [12]. Measurement systems also include software for collecting the measured data and translating the desired values to STC with existing I-V curve with given parameter coefficients.

It is well known that PV modules operate under a wide range of temperature, irradiation level, angles of incidence of the sunlight, and spectral distribution. All these conditions affect the electricity output of the PV module. The temperature dependency of the PV module can be determined from the I-V curves at different temperature values and constant irradiance values in laboratory tests. In a similar line with these temperature dependencies, irradiance dependency, spectral response, and thermal behavior characteristics can be determined. There are a lot of universities and research labs all over the world that use several setups for measuring I-V at indoor, but it is not so easy to control and arrange some outdoor parameters. Generally, some meteorological parameters are used in the calculations and measured in different setups, or they can be included in the I-V curve measurement system shown in **Figure 5**.

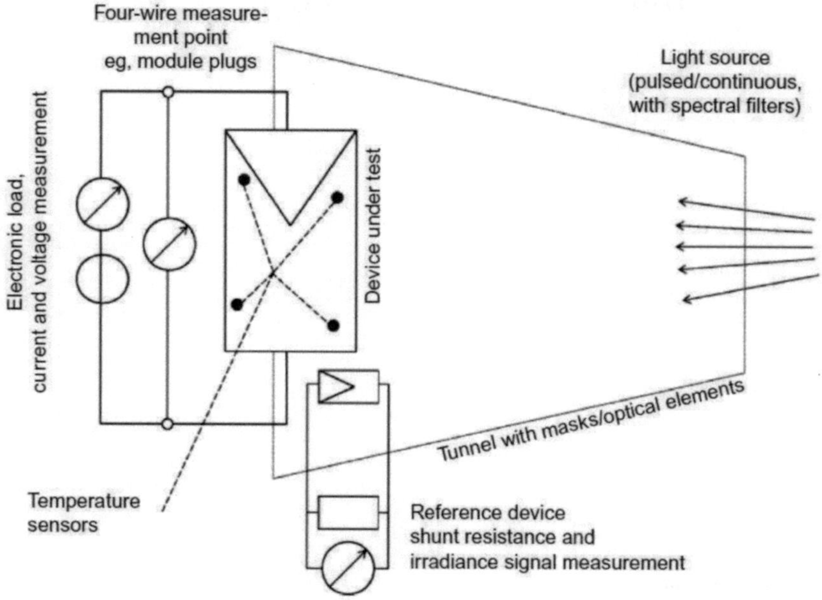

Figure 4.
Schematic setup for I-V curve measurements at laboratory [13].

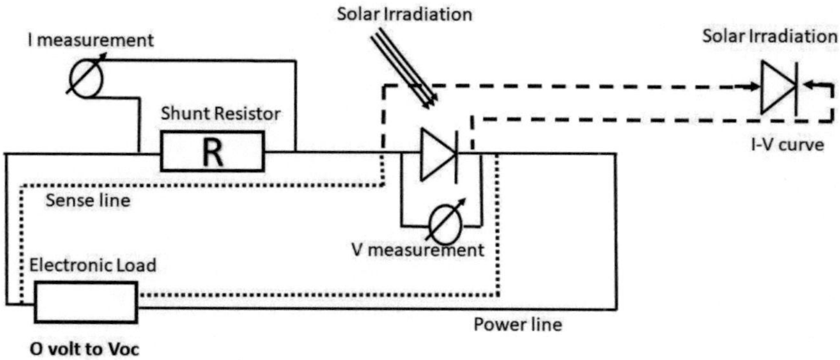

Figure 5.
Schematic setup for I-V curve measurements at outside.

In some measurements, sense measurements are not used, but the presence of sense measurements supports the accuracy of the collected data. And, current measurements are evaluated with a precisely known value of shunt resistors. These sets of measurements increase the accuracy of the data. It is also difficult to get data close to short circuit, because the presence of a load complicates collecting data. An external power supply helps overcoming this problem, and researchers get more precise data for drawing the I-V curve and determining the PV module parameters [14].

5. Prediction of PV yield

PV module characterization methods are outlined in Section 4. Computer simulation tools are used to predict the electricity production of PV systems which are necessary for economic decisions [15]. Some input parameters like operation situations, environmental conditions and the location of the system are

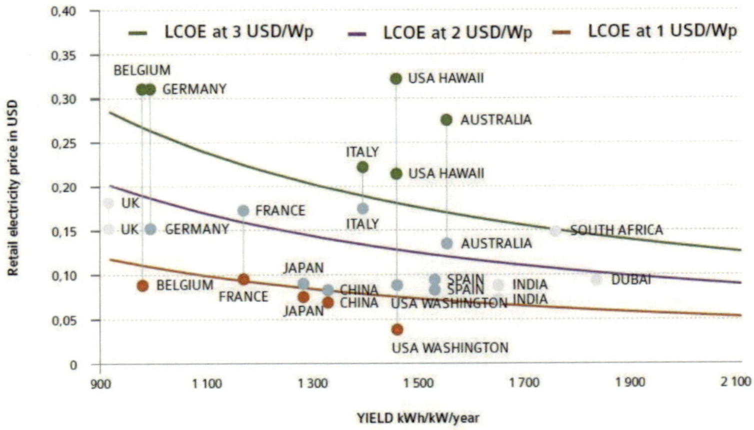

Figure 6.
PV electricity cost as a function of solar irradiance [4].

necessary for characterizing PV module and for electricity output of the system. The required number of input parameter depends on the complexity of the tool. The prediction of PV electricity yield is very important. Different technologies are used in PV systems [16–18]. Crystalline silicon technology based on PV modules dominates the market. Besides, there are other technologies which are used as PV modules that depend on the location of the installation and amount of electricity produced and meet in seasons. In sunny and coastal regions in summer, thin-film PV modules will perform better than crystalline silicon PV modules, while in high locations and during winter season crystalline silicon-based PV modules perform better than thin film because of the temperature coefficients. To understand technological differences and over or under performance of one technology with respect to other under specific climatic conditions, not only the power under STC is enough. It should be necessary to know the quality and the performances of other equipment in PV system [9]. Most of the people (users or investors) mind only the electricity yield while they decided to install a PV system. For grid-connected PV systems, electricity cost (levelized cost of electricity (LCOE)) has to be comparable with grid prices. For increasing the PV electricity usage or producing PV system equipment, there are different support schemes in countries. This support appears sometimes in tax incentives sometimes like feed in tariff (FiT) and sometimes in self-consumption or various schemes. Grid parity is caught for several countries which have higher annual irradiation and sometimes different supporting schemes like self-consumption are applied. For the other countries, there are different support schemes for declining electricity cost. LCOE as a function of solar resource for some countries are given in **Figure 6** [4]. Installation cost per kW-dependent LCOE is also given in **Figure 6**. LCOE depends on PV system size and location, so retail prices for some countries are wide (e.g., USA; although PV system has the same size, electricity prices are in the range of 0.04–0.32 USD).

6. Conclusions

In this chapter, PV electricity and the characterization methods used to determine PV module are summarized. PV performance measurement methods and electricity cost per installed power is discussed. Although installed PV capacity on earth is about 500 GW, PV contribution to global electricity demand is still less than

3%, PV installation cost continues to decrease, and the main point is the cost of PV electricity. In order to be competitive with the market prices, PV electricity cost will have to be as low as it. This will be achieved with large-scale PV installations, low installation costs, and low maintenance cost.

Author details

Rustu Eke
Department of Physics, Faculty of Sciences, Mugla Sıtkı Kocman University, Mugla, Turkey

*Address all correspondence to: erustu@mu.edu.tr

References

[1] Haque A, Bharath KVS, Khan MA, Khan I, Jaffery ZA. Fault diagnosis of photovoltaic modules. Energy Science & Engineering. 2019;7:622-644. DOI: 10.1002/ese3.255

[2] Jäger-Waldau A. Snapshot of photovoltaics—February 2019. Energies. 2019;**12**:769. DOI: 10.3390/en12050769

[3] McAvoy A, Markvart T, Castaner L. Practical Handbook of Photovoltaics Fundamentals and Applications. 2nd ed. Oxford: Elsevier; 2012. 1244p. ISBN: 9780123859341

[4] IEA Photovoltaic Power System Programme. Trends 2018 in photovoltaic applications. Report No. IEA-PVPS T1-34; 2018. 86p. ISBN: 978-0-906042-79-4

[5] Sundaram S, Benson D, Mallick TK. Solar Photovoltaic Technology Production. 1st ed. UK and USA: Elsevier; 2016. 116p. DOI: 10.1016/C2014-0-04034-8

[6] Green MA, Hishikawa Y, Dunlop ED, Levi DH, Hohl-Ebinger J, Yoshita M, et al. Solar cell efficiency tables (Version 53). Progress in Photovoltaics. 2019;**27**:3-12. DOI: 10.1002/pip.2102

[7] Luque A, Hegedus S, editors. Handbook of Photovoltaic Science and Engineering. Chichester: Wiley; 2003. 1138p. ISBN: 0471491969

[8] Pearsall N, editor. The Performance of Photovoltaic (PV) Systems. Cambridge: Elsevier; 2017. 1348p. ISBN: 978-1-78242-354-6

[9] IEA Photovoltaic Power System Programme. Photovoltaic module energy measurements: Existing approaches and best practice. Report No. IEA-PVPS T13-11; 2018. 134p. ISBN: 978-3-906042-52-7

[10] IEC 61853-1. Photovoltaic Module Performance Testing and Energy Rating—Part 1 [Internet]. 2011. Available from: https://webstore.iec.ch/publication/6035 [Accessed: April 1, 2019]

[11] Eke R, Demircan H. Performance analysis of a multicrystalline Si photovoltaic module under Mugla climatic conditions in Turkey. Energy Conversion and Management. 2013;**65**:580-586. DOI: 10.1016/j.enconman.2012.09.007

[12] Makrides G, Phinikarides A, Sutterlueti J, Ransome S, Georghiou GE. Advanced performance monitoring system for improved reliability and optimized levelized cost of electricity. In: Proceedings of 32nd European Photovoltaic Solar Energy Conference and Exhibition; 20-24 June, 2016; Munich, Germany; 2016. pp. 1973-1977

[13] Dirnberger D. Photovoltaic module measurement and characterization in the laboratory. In: Pearsall N, editor. The Performance of Photovoltaic (PV) Systems. Cambridge: Woodhead publishing; 2017. 1348p. ISBN: 978-1-78242-354-6

[14] Eke R. Correlation of operating conditions and solar cell parameters [Thesis]. İzmir: Ege University; 2007

[15] Fanney AH, Davis MW, King DL, Boyson WE, Kratochvil JA. Comparison of photovoltaic module performance measurements. Journal of Solar Energy Engineering. 2006;**128**(2):152-159. DOI: 10.1115/1.2192559

[16] Kenny RP, Ionnides A, Müllejans H, Zaaiman W, Dunlop ED. Performance of thin film PV modules. Thin Solid Films. 2016;**511-512**:663-672. DOI: 10.1016/j.tsf.2005.11.066

[17] Carr AJ, Pryor TL. A comparison of the performance of different PV module

types in temperate climates. Solar Energy. 2004;**76**(1-3):285-294. DOI: 10.1016/j.solener.2003.07.026

[18] Notton G, Lazarov V, Stoyanov L. Optimal sizing of a grid-connected PV system for various PV module technologies and inclinations, inverter efficiency characteristics and locations. Renewable Energy. 2010;**35**:541-554. DOI: 10.1016/j.renene.2009.07.013

Chapter 2

Comparing Different PV Module Types and Brands Under Working Conditions in the United Kingdom

Mohamad Kharseh and Holger Wallbaum

Abstract

The present work demonstrates the performance evaluation and economic analysis of different PV module types and brands at the working conditions of Padiham (53.5 N, 2.3 W) in the UK. The total area of PV plant was assumed to be 100 square meters. The simulations were carried out for modules installed on the roof and on the south-facing façade of a residential building. The comparison study is carried out to define the most suitable module type and brands for the considered place in the current study. The energy and economic performance of the grid-connected PV system are analyzed under the meteorological conditions of Padiham. The modules were characterized by evaluating their annual electrical energy generation and different figures of merit of the grid-connected PV systems such as the investment, annual profit, net present value, levelized cost of electricity, and the payback time. The simulations show that in this specific setup, monocrystalline modules have the best energy performance, while thin-film modules have the best economic performance.

Keywords: PV system, simulations, different PV brands, potential

1. Introduction

Today, most of the global energy comes from using fossil energy carrier. There is mounting evidence that global warming and, consequently, climate changes are anthropogenic and attributed to fossil-fuel consumption. The strong fossil-fuel dependence of our current energy system has resulted in daily fossil-fuel consumption of 98.2 million barrels of oil equivalent (Mbbl) in 2017 [1]. It is worth to mention that fossil fuel represents 75% of total global energy consumption, while renewable energy resources represent only 25% [2]. Consequently, about 36.8 billion tons of carbon dioxide emissions are emitted into the atmosphere annually for global energy production [3]. Observations provide evidence that atmospheric CO_2 levels caused by human activities have increased by 25% over the last century, which leads to rising global temperature [4].

In helping to face those challenges and achieving the EU targets for 2020 and 2050, in addition to improving the performance of the existing energy conversion systems, utilizing the available local renewable energy resources is needed as well. Although the current efficiency of such systems is still relatively low and the capital cost is still high, the abundance of solar energy that strikes our planet makes the use of solar panel economically viable [5, 6]. It is worth mentioning that the growing demand for renewable energy sources in recent years led to advancing the manufacturing of solar cells. The fast decrease in system prices combined with the increases in the performance make it economically feasible, and, consequently, increased amount installed globally [7]. Regarding the worldwide installed capacity, solar photovoltaic is now the third largest renewable energy source with annual capacity additions reaching 93.7 GW and 94.3 GW in 2017 and 2018, respectively [2, 8]. By the end of 2018, the total accumulated capacity is 480 GW [8]. Because they have no moving parts, PVs are stable over time, with typical durability of 25 years, and low maintenance is required during their operation [9].

Indeed, many different PV module types and brands are available in the market. The performance of each type and brand varies from one to another and strongly depends on the meteorological and working conditions. Current work, therefore, presents a comparison between different module types and brands under the meteorological conditions of Padiham (53.5 N, 2.3 W) in the UK.

2. Photovoltaic system overview

Photovoltaic (PV) system is a technology that converts solar energy into electricity. Because no moving parts, such systems are reliable, and low maintenance is required during the operation. In addition to the fact that PV systems do not emit any greenhouse gases during operation, manufacturing such systems results in very low emissions. In average, less than 2 years is needed for a PV module to generate the same amount of energy that has been used to fabricate the module—depending on the type of module [10].

Recently, the PV module has been integrated to building external envelope. Such a system is the so-called building-integrated photovoltaics (BIPV). The most advantage of BIPV is that such a system generates the electricity at the place where it is demanded, as well as PV modules can replace some parts of the building envelope components. This can result in reducing the initial costs of BIPV due to lowering the costs of conventional envelope materials. PV modules can be integrated into different parts of the envelope such as roof (flat or slope), façade, and shading devices for windows.

The electric output of a PV system depends on different factors such as:

- The availability of and accessibility to solar radiation, which in turn is influenced by the climate, the inclination, and orientation of the modules
- The PV technology is related to facts such as efficiency and its decline with time and the cell temperature
- The over shading in some areas of the modules

3. Objective

This work aim is to evaluate the performance of a photovoltaic (PV) system installed on the roof or integrated to the south-facing façade of residential apartments in Padiham (53.5 N, 2.3 E), UK. Moreover, the work presents a comparison

	Type		
Brands	Monocrystalline	Polycrystalline	Thin film
	Motech XS72D3-320	Hanwha Solar One 310	Stion STO-150
	Panasonic N330	JA Solar 48-225/4BB	Solar Frontier SF170-S
	Silevo 310	Trina 315	Stion STN-125
	Sunpreme 360	SunTech 320	SolTech Energy 80
	Solar World SW350 XL	Jinko Solar JKM270	
	Silevo 220	Motech IM72C3-310	
	Optimus 60 OPT280	Canadian Solar CS6X-320P	
	LG Solar 320	Solar World SW260	
	Jinko Solar JKM275M	Yingli Solar YL305P	
	Solar World SW325 XL	PEAK BLK 255	
	JA Solar 72-315/SI	Gintech 240	
	Hyundai HiS-S265MG	Hyundai HiS-S250MG	
	Mitsubishi PV-MLE265HD2	Kyocera KD220GH-4FB2	

Table 1.
Considered PV module types and brands [11].

between the energy and economic performance of the most common PV module types, which are:

1. Polycrystalline module

2. Monocrystalline module

3. Thin-film module

Of course, a tremendous number of companies are making different brands of PV modules. Therefore, in order to identify the suitable module type and brand, performances of different module type and brands are compared under the meteorological conditions of Padiham, UK. The present work considers only the brands that were mentioned in the report prepared by the Fraunhofer Institute for Solar Energy Systems, ISE [11] (see **Table 1**). The electricity generation profiles of the installed PV system will be identified. Annual electricity generation profile, energy efficiency, and different figures of merit of economic performance of different module types and brands will be illustrated at the working conditions of the considered city.

4. Determination of available solar energy

The first and most important step before designing a photovoltaic system is the determination of the available solar energy (ASE) on an inclined surface in the considered site. ASE can be either measured or simulated. Solar radiation simulations have advantages over measurements and are more reliable over the years [12]. Unlike measured solar radiation, simulated solar radiation can account for universal climate variations over many years, without having the burden of having to process decades of field data. Also, the actual measurements of the ASE are

January	February	March	April	May	June	July	August	September	October	November	December
0.32	0.34	0.37	0.39	0.41	0.39	0.41	0.41	0.38	0.34	0.30	0.30

Table 2.
Clearness index in Padiham, UK [18].

Input Data

Latitude	55.871254	N	degree
Longitude	12.826451	e	degree
Standard Longitude	15	degrees	
Slop angle β	50.0	degrees	
Surface azimuth	180.0	degrees	
Avilable area	100	m^2	
Inflation rate	0.7%	%	
Real interest rate	3.4%		
Electrical price escalation rate	1.0%		
Electricity price	0.207	$/kWh	

Month	Clearness KT
Jan	0.33
Feb	0.39
Mar	0.44
Apr	0.48
May	0.5
Jun	0.48
Jul	0.49
Aug	0.49
Sep	0.45
Oct	0.38
Nov	0.39
Dec	0.35

PV Specification		
fill factor	0.77	-
derating factor	0.90	
short circuit current	9.09	A
open circuit voltage	45.73	V
voltage temp. coefficient	340.00	%/C
NOCT (nominal operating cell temperature)	45.00	C
nomional capacity	320.00	W
module price	480.00	$
Area	1.64	m^2
module degradation rate	1.0%	%

Figure 1.
Input data interface of the developed excel-based model to simulate the PV system.

costly due to the high price of the required instruments. Therefore, simulation is a common method to calculate the available solar energy at a particular location. In this work, a computational model was built to estimate the available solar energy, with the resolution of 1 hour, per square meter of surface considering different slopes and azimuth angles. The model can be used to determine the optimal azimuth and slope angles of the PV modules. The optimum angles are defined as the angles that result in maximum annual electricity generation. More details about the model can be found in Ref [13].

It should be noted that the model was built based on well-known theoretical relations for calculating available solar energy. These relations have been tested by a number of studies [14, 15]; the reader is recommended to consult Section 2 in Refs. [16, 17]. It is also worth noting that in the current study, the cloudy day approach was taken in the calculation of the available solar energy. The clearness index of the city of Padiham, which was taken from [18], is tabulated in **Table 2**. The specifications of the considered PV module, which appear in **Figure 1** as input data, were collected from the technical brochure of each module, which was provided by the factory of each considered module. Some of the specifications of the PV module are certain wattage, voltage, and amperage which are called electrical characteristics of the module. These characteristics of the module are defined under specific conditions which are called standard test conditions (STC). STC, which is the universal standard, is a set of laboratory test conditions including irradiance of 1000 W/m^2, air mass of AM1.5, and cell temperature of 25°C [16].

5. Simulation PV system

Because an electric grid is available in Padiham, a grid-connected system will be considered in the current work. In such a system, the electric grid acts as a virtual storage system, which in turn can reduce the net costs of PV systems. However, these systems require an inverter to convert the DC (direct current) voltage into AC voltage. Also, to reduce the initial cost of the system, grid-connected PV systems have another advantage. Indeed, the electricity consumption of a building is not constant over the day and varies depending on the occupancy of the building, user behavior, and used equipment. On another side, the electricity output of the PV system varies during the day with the position of the sun in the sky.

Consequently, there might be periods when the output of the PV system exceeds the electricity demand of the building or vice versa. Therefore, in a grid-connected system, the electricity demand of the building is met by a combination of solar energy and grid electricity. In other words, when the output of the PV system exceeds the demand of the building, the system will export the exceeded energy to the grid and vice versa. Namely, when the output of the PV system is not enough to cover the demand of the building, the system will import electricity from the grid.

In current work, an Excel-based model has been built to simulate hour-by-hour energy and economic performance of a grid-connected PV system for different slope angle. As mentioned above the model can be used to determine the optimal inclination angle so that the annual generation of PV module can be maximized. The model can also simulate a PV system installed on south-facing façade or a vertical wall of any azimuth angle. Furthermore, the model was built using well-known theoretical relations for calculating the electricity generation of the PV system. For more details, the reader is recommended to consult Refs. [19, 20]. In order to be able to simulate the energy and economic performance of the PV system, assumptions had to be made. In the current study, the assumptions are based on realistic current conditions in the UK and data collected from literature review (see **Table 3**)[1] [21–33].

5.1 Component and installation costs

The price PV module was taken based on the price given by different suppliers. According to the component price collected from literature review, the replacement of the inverter (which takes place every 8th year) and the installation cost were assumed to be 322 $/kW and 0.018 $/W, respectively [23–25].

5.2 Operation and maintenance costs

Operation and maintenance (O&M) represent expenses on equipment and services that occur after the system is installed. Fortunately, solar panels have 25 years

Factor	Value	Factor	Value
Inflation rate	0.1%	Labor cost	18 $/kW
Real interest rate	0.6%	Operation	10 $/kW·y
Elect. price	20.1 ₵/kWh	Inverter lifetime	8 y
Feed-in tariff	6.7 ₵/kWh	Wire efficiency	98%
Elect. price escalation	1%/y	Lifetime of the project	25 y
Degradation rate	1%/y	Inverter price	322 $/kW
Cell temperature	$T_C = T_a + 1.25 \bullet (NOCT - 20) \bullet G_s$		

Table 3.
Assumptions made in the present work.

[1] T_c is the cell temperature, T_a is the ambient air temperature, NOCT is the nominal operating temperature, and Gs is the available solar radiation (W/m^2).

of warranty. Product warranties usually cover major maintenance. Therefore, O&M costs during the first 25 years (which was assumed to be the project economic life) will be only for surface cleaning of PV modules and supervising. In 2013 in the USA, the cost of supervision and twice a year cleaning of the panels was reported to be 8.3 $/kW per year [22]. Therefore, in current work, the annual O&M cost of the project will be assumed 10 $/kW·y.

6. Results and discussions

The developed model was used to simulate the performance of grid-connected PV system. As mentioned above, different brands of PV modules were tested in the present work, see **Table 1**.

6.1 Thin-film module type

Among the brands of the thin film mentioned in the report prepared by the Fraunhofer Institute for Solar Energy Systems, ISE [11], only the price of four brands was collected so far.

The electrical output of the PV module per square meter of different brands of thin-film PV module installed on the roof is shown in **Figure 2**. As shown, the annual electrical output per square meter of selected modules varies from a minimum of 127 to a maximum of 151 kWh/m2·y. The optimal slope angle of the module was found to be 49° facing the south. From an energy viewpoint, the results illustrated in **Figure 2** show that Stion STO-150 thin film is the best among the considered thin-film module brands.

Energy performance of PV modules strongly depends on the module temperature variations. Therefore, as expected, simulations show that the efficiency of the PV modules during colder months is higher than those during the warmer months. **Figure 3** shows that the efficiency of the thin-film PV module ranged from 9.6% (SolTech Energy 80 in August) to 12% (Stion STO-150 in January).

In order to compare the different brands from the economic point of view, the levelized cost of electricity and the accumulative cash flow were determined for each brand. As shown in **Figure 4**, the levelized cost of electricity varies from a

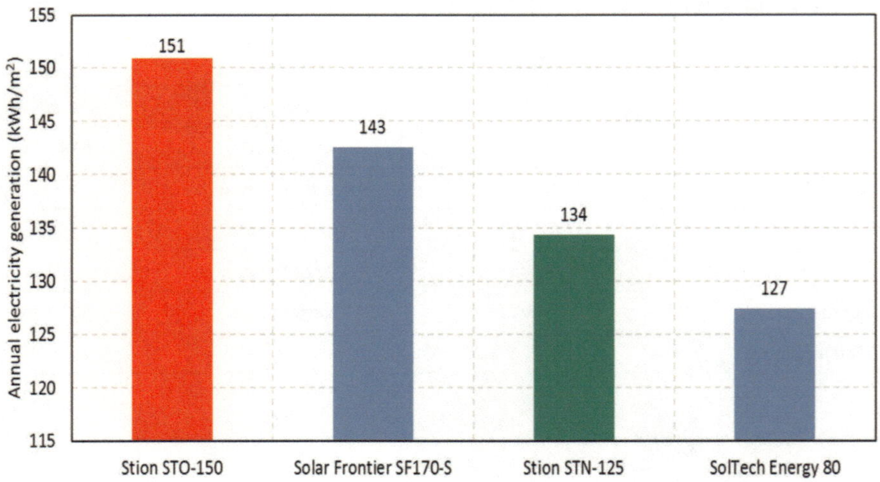

Figure 2.
Electricity output per square meter of thin-film PV module installed on the roof. The optimal slope angle of the module was found to be 49 facing the south.

minimum of 7.5 (Stion STN-125) to a maximum of 15.3 ₵/kWh (SolTech Energy 80). Thus, from the levelized cost of electricity viewpoint, Stion STN-125 is the best among the considered brands.

Figure 5 shows the accumulative cash flow of Stion STO-150 (of the best energy performance) and Stion STN-125 (of the best economic performance). Apparently, the installation cost of Stion STO-150 is 39% more than the initial cost of Stion STN-125 for the same available area of PV cells (which is in our case 100 m^2).

In other words, by using Stion STO-150, the annual electricity is increased by 12%, and the initial cost of such a system is increased by 39% in comparing to Stion STN-125 (see **Table 4**). Hence, the preference of the decision-maker should be considered in order to select the most suitable PV brand, namely, either the brand of the best energy performance or the brand of the best economic performance.

Repeating the simulation for the module installed on the south-facing façade leads to the same conclusion. Explicitly, from an energy viewpoint, the Stion STO-150 thin film is the best brand among the considered brands. While from the economic viewpoint, Stion STO-125 thin film is the best brand among the considered brands.

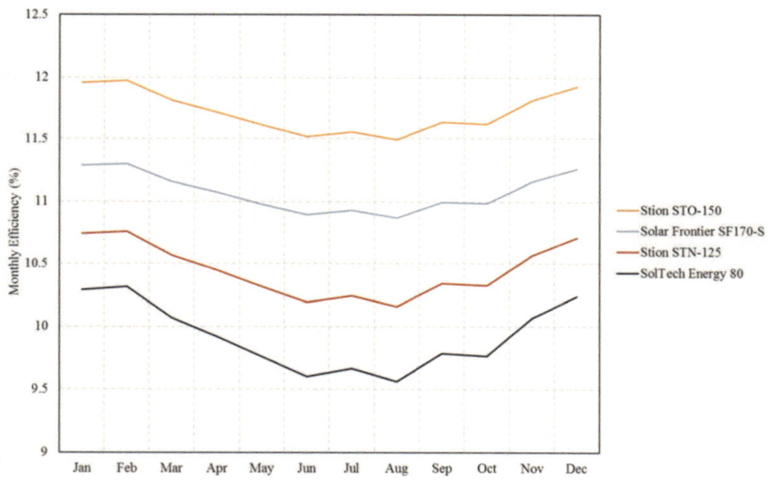

Figure 3.
Monthly average efficiency of different brands of thin-film module at working conditions of Padiham, UK.

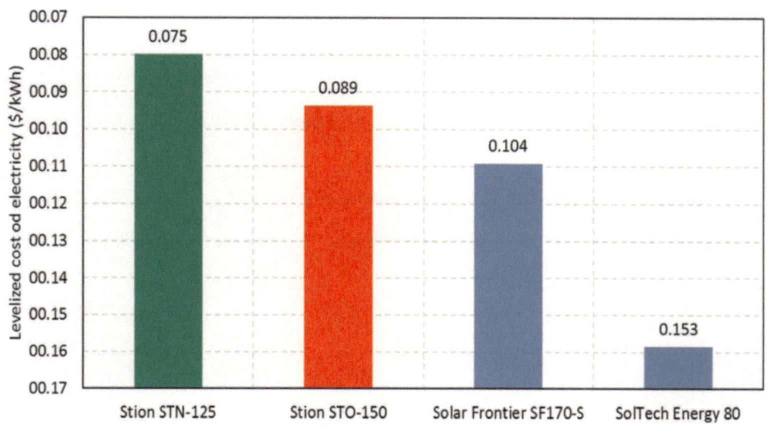

Figure 4.
The levelized cost of electricity of different brands of thin-film module at working conditions of Padiham, UK.

Table 5 shows the comparison between two different thin-film PV module brands that are installed on the roof of the building and installed on the southern façade. As shown, modules that are installed on the roof can generate up to 31% more annual electricity as compared with the same modules installed on the façade. This can be attributed to the fact that a module on the rood receives more solar energy than a module installed on the façade. On the other hand, the greater solar radiation on the roof makes the temperature of modules that are installed on the roof slightly higher. Consequently, the average annual efficiency of the system installed on the façade will be slightly higher. However, the abundant amount of the solar energy on the roof, as compared with the façade, overcomes the improvement in the efficiency. So, as it was expected, installing a PV system on the roof is better than on the southern wall.

6.2 Polycrystalline module type

In this section, a comparison between 13 different brands of the polycrystalline PV module is presented. Recall that the selected brands were chosen based on the report prepared by Fraunhofer Institute for Solar Energy Systems, ISE [5].

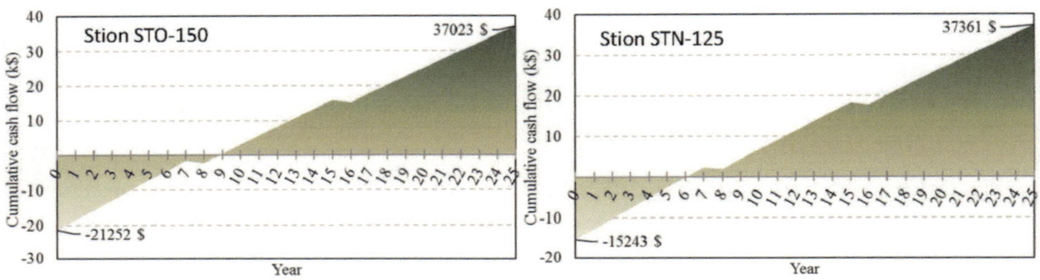

Figure 5.
Accumulative cash flow of Stion STO-150 (on the left) and Stion STN-125 (on the right) at the current conditions in the UK.

Brand	Annual elect. (kWh/m^2)	Annual efficiency (%)	LCOE (₵/kWh)	Initial cost ($/100 m^2)	Payback time (year)
Stion STO-150	151	11.7	8.9	21,252	9
Stion STN-125	134	10.4	7.5	15,243	6
Difference	12%	13%	19%	39%	50%

Table 4.
Comparing two thin-film modules of best energy and best economic performance.

	Annual elect. (kWh/m^2)			Avg. annual efficiency (%)			LCOE (₵/kWh)		
	Roof	Facade	Difference (%)	Roof	Facade	Difference (%)	Roof	Facade	Difference (%)
Stion STO-150	151	115	31	11.7	11.8	−1	8.9	11.6	23
Stion STN-125	134	103	30	10.4	10.6	−2	7.5	9.8	23

Table 5.
Comparison between installing modules on the roof and on the south-facing façade.

The annual electrical output per square meter of different brands of polycrystalline that are installed on the roof is illustrated in **Figure 6**. As shown, the annual electrical output per square meter of selected modules varies from a minimum of 162 to a maximum of 215 kWh/m^2 y. The optimal slope angle of the module was found to be 48° facing south.

As expected, simulations show that the efficiency of the polycrystalline PV modules during colder months is higher than those during the warmer months. The PV module voltage reduction causes efficiency reduction due to increasing the operating temperature of the module. **Figure 7** shows that the efficiency of polycrystalline PV module ranged from 11.7% (Kyocera KD220GH-4FB2 in August) to 17.6% (Hanwha Solar One 310 in February).

Hence, from the energy viewpoint, Hanwha Solar One 310 module seems to be the best among the considered polycrystalline module brands.

In order to compare the different brands from the economic point of view, the levelized cost of electricity and the accumulative cash flow were determined for

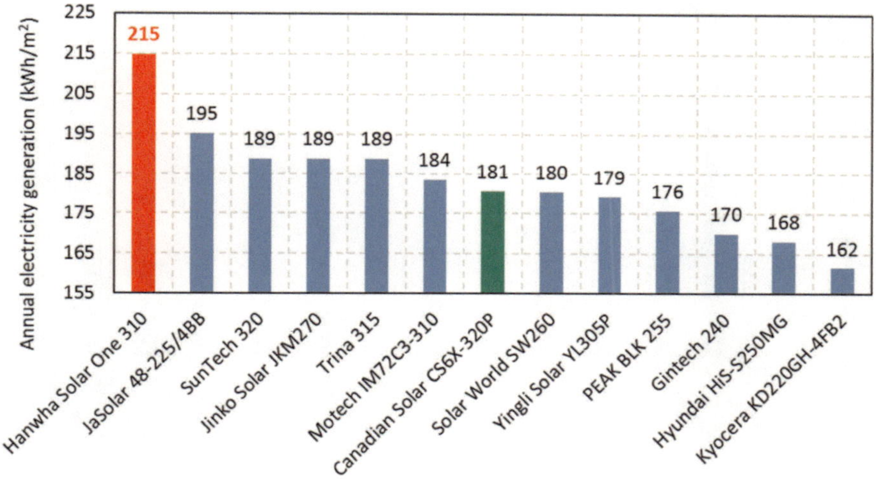

Figure 6.
Electrical output per square meter of polycrystalline PV module installed on the roof. The optimal slope angle of the module was found to be 48 facing south.

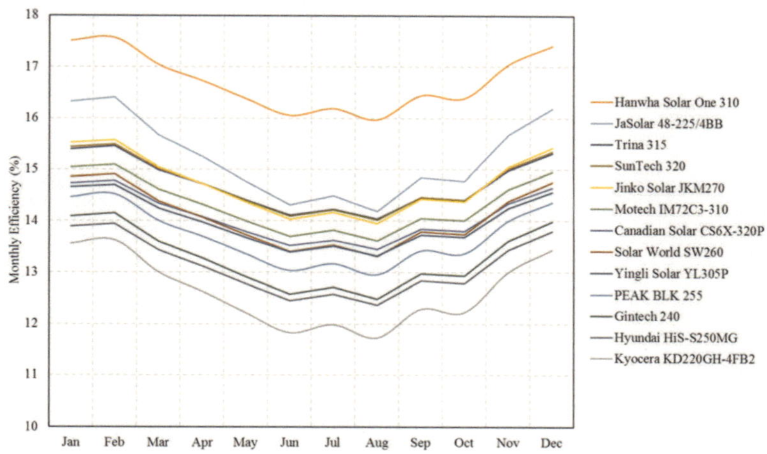

Figure 7.
Monthly average efficiency of different brands of the polycrystalline module at working conditions of Padiham, UK.

each brand taking into account the assumptions shown in **Table 3**. As shown in **Figure 8**, the levelized cost of electricity varies from a minimum of 7.8 (Canadian Solar CS6X-320P) to a maximum of 12.1 ₵/kWh (Kyocera KD220GH-4FB2).

Figure 9 shows the levelized cost of electricity versus the annual electricity generation per square meter of different polycrystalline brands. One can say that the annual electricity generation and levelized cost of electricity are somehow contradicting. As shown the Hanwha Solar One 310 seems to be the best from annual electricity generation viewpoint, while Canadian Solar CS6X-320P seems to be the best from levelized cost of electricity viewpoint.

Figure 10 shows the accumulative cash flow of Hanwha Solar One 310 (of the best energy performance) and Canadian Solar CS6X-320P (of the best economic performance). Apparently, the installation cost of Hanwha Solar One 310 is 61% more than the initial cost of Canadian Solar CS6X-320P for the same available area of PV cells (which is in our case assumed 100 m^2). In another word, by using Hanwha Solar One 310, the annual electricity is increased by 19%, and the initial cost of such system is increased by 61% in comparison to Canadian Solar

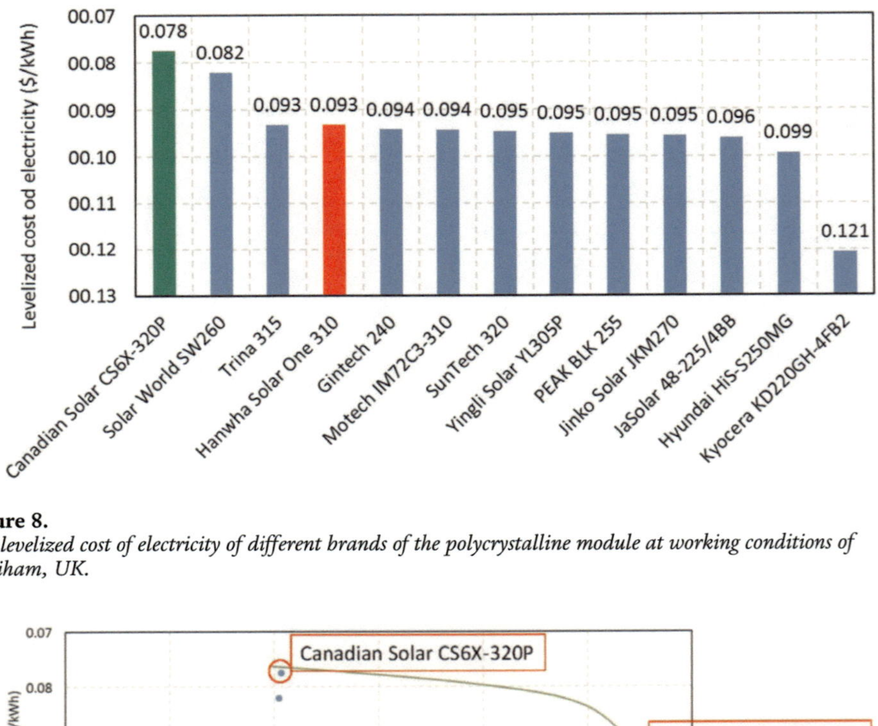

Figure 8.
The levelized cost of electricity of different brands of the polycrystalline module at working conditions of Padiham, UK.

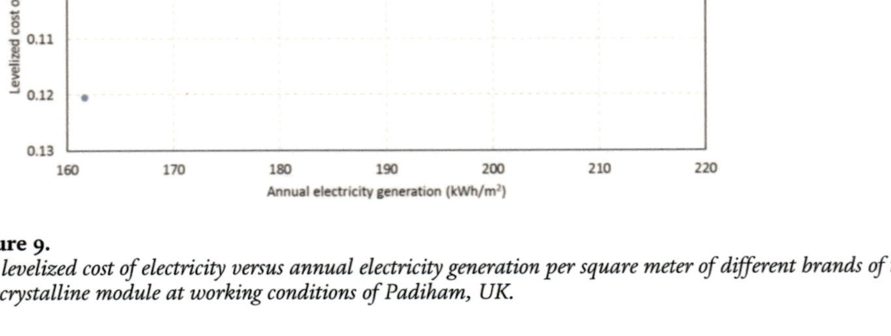

Figure 9.
The levelized cost of electricity versus annual electricity generation per square meter of different brands of the polycrystalline module at working conditions of Padiham, UK.

CS6X-320P (see **Table 6**). Hence, the preference of the decision-maker should be considered in order to select a suitable PV brand.

In order to investigate the potential of installing polycrystalline PV modules on the south-facing façade, the analyses were repeated for the modules installed on south-facing façade. The simulations show that a similar conclusion can be drawn with regard to modules installed on the roof. Namely, in the case of installing PV cell on the south-facing façade, among the considered brand (see **Table 1**), the best polycrystalline brand from energy viewpoint is Hanwha Solar One 310, while Canadian Solar CS6X-320P seems to be the best brand from the economic viewpoint. **Table 7** shows the comparison between the two different polycrystalline PV module brands installed on the roof of the building and installed on the south-facing façade. As shown, modules which are installed on the roof can generate 29–30% more electricity per year. This is due to the fact that the roof module receives more solar energy compared to the façade. Still, the temperature of modules installed on

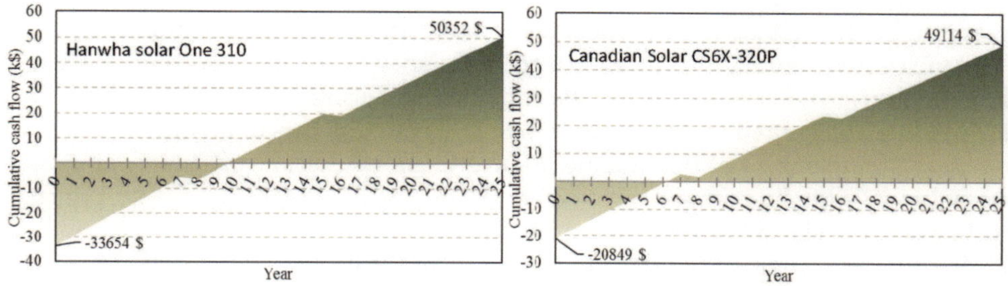

Figure 10.
Accumulative cash flow of Hanwha solar one 310 (on the left) and Canadian solar CS6X-320P (on the right) at the current conditions in the UK.

	Annual electricity (kWh/m^2)	Avg. annual efficiency (%)	LCOE (₵/kWh)	Initial cost ($/100 m^2)	Payback time (year)
Hanwha Solar One 310	215	16.7	9.3	33,654	9.8
Canadian Solar CS6X-320P	181	14.0	7.8	20,849	6.2
difference	19%	19%	19%	61%	58%

Table 6.
Comparing two polycrystalline modules of best energy and economic performance.

	Annual elect. (kWh/m^2)			Avg. annual efficiency (%)			LCOE (₵/kWh)		
	Roof	Facade	Difference (%)	Roof	Facade	Difference (%)	Roof	Facade	Difference (%)
Hanwha Solar One 310	215	166	30	16.7	17	−2	9.3	12.1	−23
Canadian Solar CS6X-320P	181	140	29	14.0	14.3	−2	7.8	10.0	−22

Table 7.
Comparison between installing modules on the roof and on the south-facing façade.

the roof will be slightly higher; the average annual efficiency of the system installed on façade will be slightly higher (**Table 7**). However, the improvement in efficiency due to the temperature effect on the façade system can be neglected as compared to the increased electricity output of the rooftop system.

6.3 Monocrystalline module type

In this section a comparison between 13 different brands of monocrystalline mentioned in the report prepared by Fraunhofer Institute for Solar Energy Systems, ISE, is shown [5].

Figure 11 illustrates the annual electrical output per square meter of different brands of monocrystalline installed on the roof. As shown, the annual electrical output per square meter of the selected modules varies from a minimum of 175 to a maximum of 226 kWh/m^2 y. The optimal slope angle of the module was found to be 48° facing the south. Thus, from the energy viewpoint, Motech XS72D3-320 module seems to be the best among the considered monocrystalline module brands.

As expected, simulations show that the efficiency of the monocrystalline PV modules during colder months is higher than those during the warmer months. The PV module voltage reduction causes efficiency reduction due to increasing the operating temperature of the module. **Figure 12** shows that the efficiency of monocrystalline PV module ranged from 12.9% (Mitsubishi PV-MLE265HD2 in August) to 18.6% (Motech XS72D3-320 in February).

In order to compare the different brands from the economic point of view, the levelized cost of electricity was calculated for each brand. The economic analyses were carried out taking into account the assumptions shown in **Table 3**. Results illustrations in **Figure 13** shows that the levelized cost of electricity varies from a minimum of 7.5 (Solar World SW325 XL) to a maximum of 12.2 ₵/kWh (Panasonic N330).

Figure 14 shows the levelized cost of electricity versus the annual electricity generation per square meter of different monocrystalline brands. As shown, the annual electricity generation and levelized cost of electricity are somehow contradicting. Apparently, among the considered monocrystalline brands, Motech

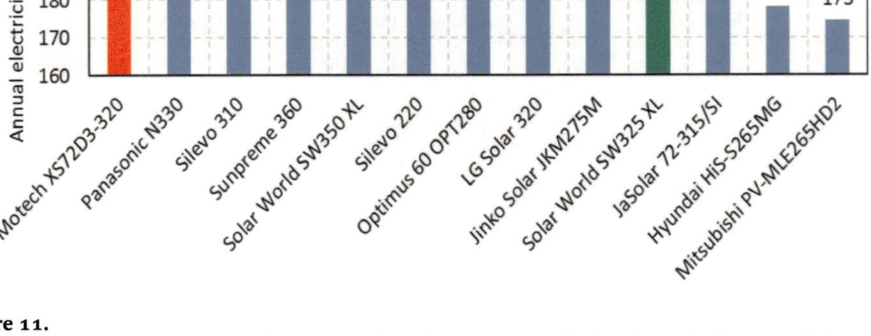

Figure 11.
Electrical output per square meter of monocrystalline PV module installed on the roof. The optimal slope angle of the module was found to be 48° facing the south.

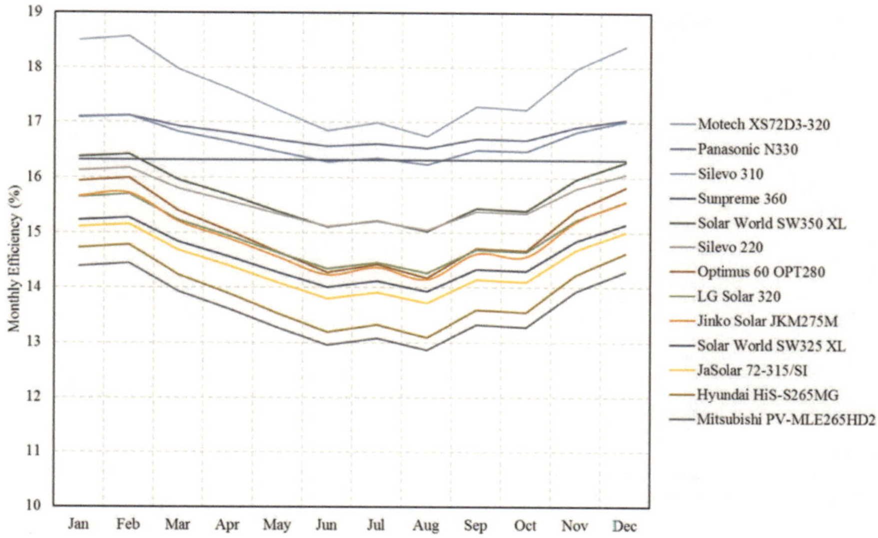

Figure 12.
Monthly average efficiency of different brands of the monocrystalline module at working conditions of Padiham, UK.

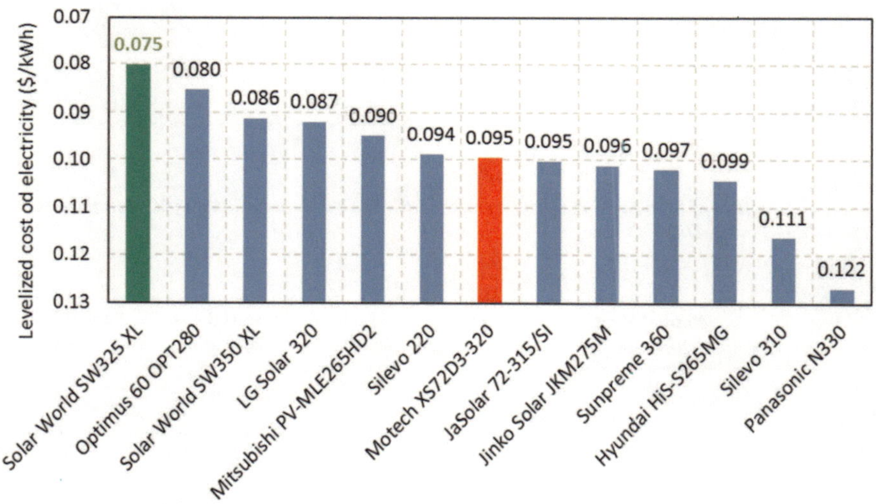

Figure 13.
The levelized cost of electricity of different brands of the monocrystalline module at working conditions of Padiham, UK.

XS72D3-320 and Solar World SW325 XL are the best from annual electricity generation and levelized cost of electricity, respectively.

Figure 15 shows the accumulative cash flow of Motech XS72D3-320 (of the best energy performance) and Solar World SW325 XL (of the best economic performance). Obviously, the installation cost of Motech XS72D3-320 is 71% more than the initial cost of Solar World SW325 XL for the same available area of PV cells (which is in our case assumed to be 100 m^2).

In another word, as shown in **Table 8**, comparing to Solar World SW325 XL, using Motech XS72D3-320 results in increasing the annual electricity by 21% (positive), while the initial cost of such system is 71% higher (negative). Hence, the preference of the decision-maker should be considered in order to choose the most suitable PV brand.

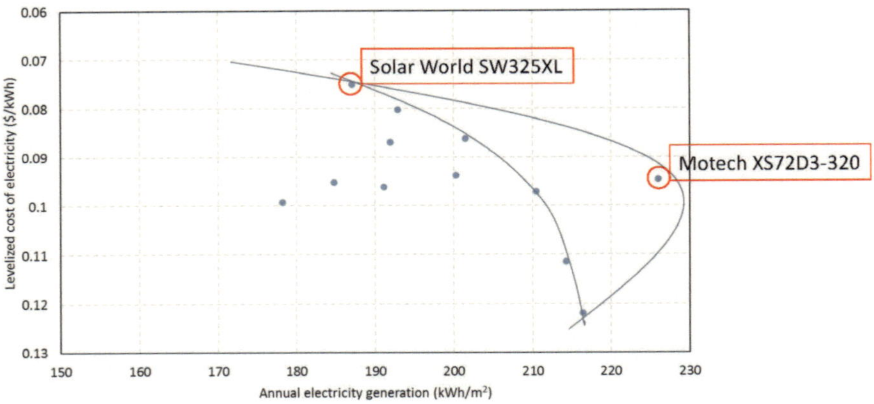

Figure 14.
The levelized cost of electricity versus annual electricity generation per square meter of different brands of the monocrystalline module at working conditions of Padiham.

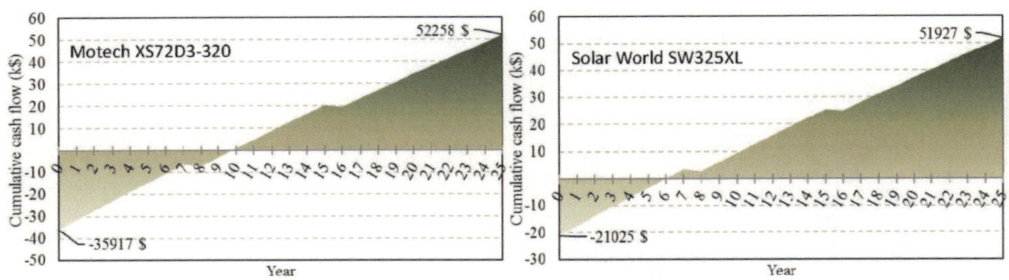

Figure 15.
Accumulative cash flow of Motech XS72D3-320 (on the left) and solar world SW325 XL (on the right) at the current conditions in the UK.

	Annual elect. (kWh/m^2)	Annual efficiency (%)	LCOE (₵/kWh)	Initial cost ($/100 m^2)	Payback time (year)
Motech XS72D3-320	226	17.5	9.5	35,917	9.9
Solar World SW325 XL	187	14.5	7.5	21,025	6
Difference	21%	21%	27%	71%	65%

Table 8.
Comparing two monocrystalline modules of best energy and economic performance.

In order to investigate the potential of installing a PV system on the south-facing façade, the analyses were repeated for the module installed on the southern façade. The simulations which show the same conclusions from the modules' installation on the roof can be drawn. Namely, in the case of installing PV cell on the south-facing façade, among the considered brands (see **Table 1**), the best monocrystalline brand from energy viewpoint is Motech XS72D3-320, while Solar World SW325 XL seems to be the best brand from the economic viewpoint. **Table 9** shows the comparison between the two different monocrystalline PV module brands installed on the roof of the building or installed on the southern façade. Because the rooftop modules receive more solar radiation than those installed on the façade, the rooftop system generates 29–30% more electricity per year. More solar energy leads to higher module temperature, and consequently, the average annual efficiency of the system installed on the façade will be slightly higher. However, the difference in efficiency can be neglected as compared to the improved electricity output of the rooftop system.

6.4 Monocrystalline vs. polycrystalline vs. thin film

Figure 16 illustrates the results of the above simulations. The brands of the best energy performance and of the best economic performance for each module type have been specified. From the energy viewpoint, Motech XS72D3-320, Hanwha

	Annual elect. (kWh/m^2)			Avg. annual efficiency (%)			LCOE (₵/kWh)		
	Roof	Facade	Difference (%)	Roof	Facade	Difference (%)	Roof	Facade	Difference (%)
Motech XS72D3-320	226	175	29	17.5	17.9	−2	9.5	12.2	−22
Solar World SW325 XL	187	144	30	14.5	14.8	−2	7.5	9.7	−23

Table 9.
Comparison between installing modules on the roof and on the south-facing façade.

PV Type	Monocrystalline		Polycrystalline		Thin film	
Brand	Motech XS72D3-320	Solar World SW325XL	Hanwha Solar One 310	Canadian Solar CS6X-320P	Stion STO-150	Stion STN-125
Energy performance	WINNER		WINNER		WINNER	
Economic performance		WINNER		WINNER		WINNER

Figure 16.
The best brands of the best economic and the best energy performance for each module type.

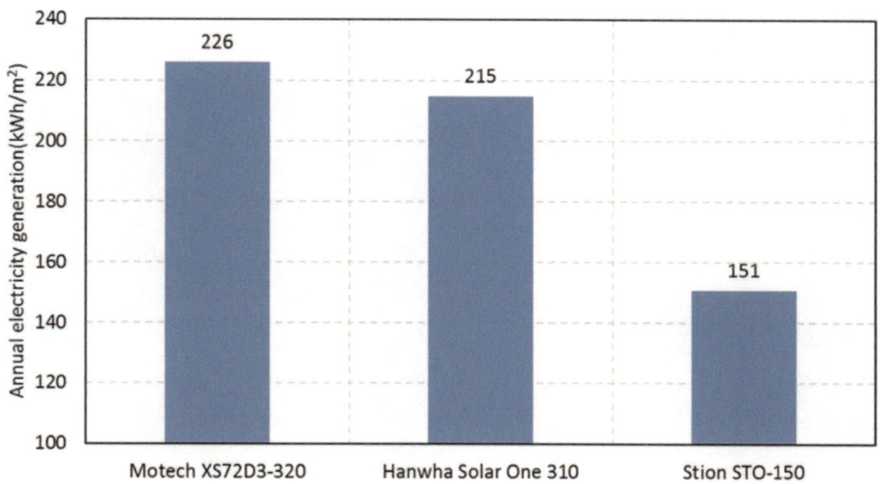

Figure 17.
Energy comparison between different modules of the best energy output.

Solar One 310, and Stion STO-150 represent the best brands for monocrystalline, polycrystalline, and thin film, respectively. While from the economic viewpoint, Solar World SW325XL, Canadian Solar CS6X-320P, and Stion STN-125 represent the best brand for monocrystalline, polycrystalline, and thin film, respectively. Hence, we do see a typical trade-off phenomenon between the environmental/energetic and the economic performance.

Figure 17 shows the comparison between electricity generation of Motech XS72D3-320, Hanwha Solar One 310, and Stion STO-150, which represent the best brands from an energy viewpoint for monocrystalline, polycrystalline, and thin film, respectively. As shown:

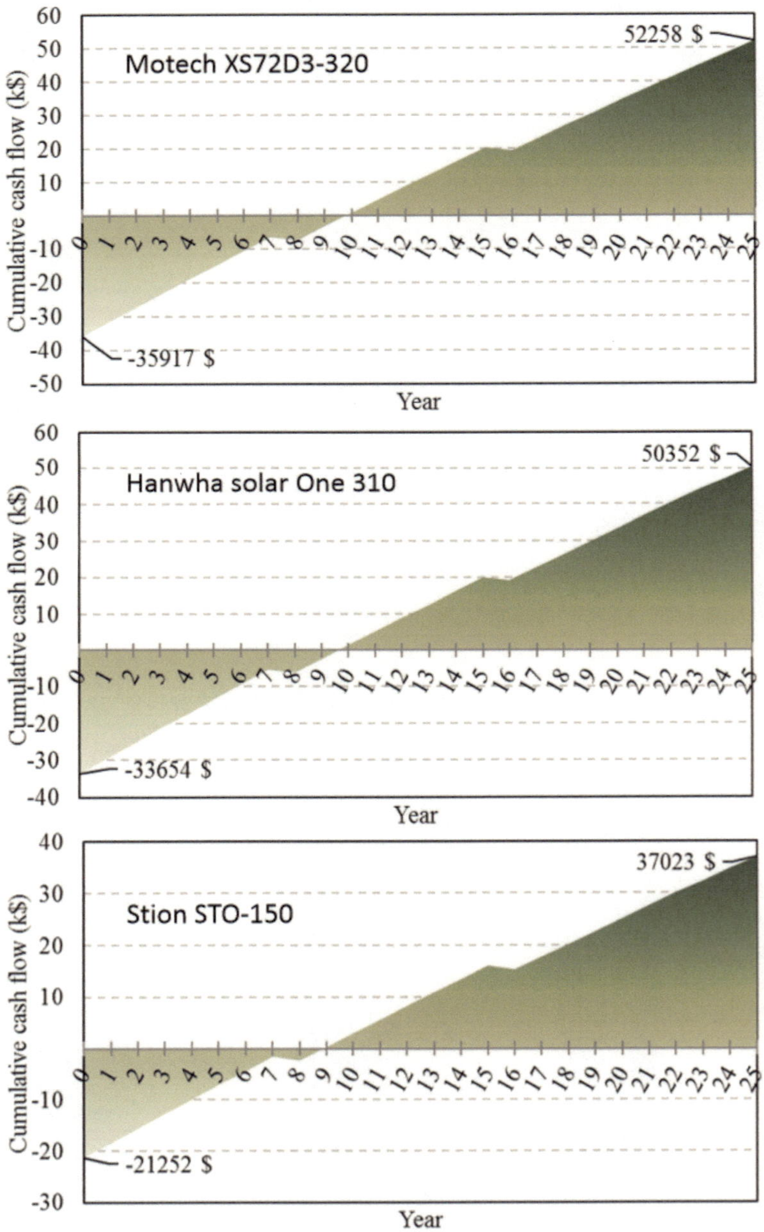

Figure 18.
Economic comparison between different modules of the best energy output.

- Monocrystalline can give 5% more annual electricity than polycrystalline.
- Monocrystalline can give 50% more annual electricity than thin-film module.
- The polycrystalline module can give 42% more annual electricity than thin-film module.

From the economic viewpoint, it seems that the thin-film module has the best economic performance (see **Figure 18**).

Figure 19 shows the comparison between the accumulative cash flow of Solar World SW325XL, Canadian Solar CS6X-320P, and Stion STN-125, which represent

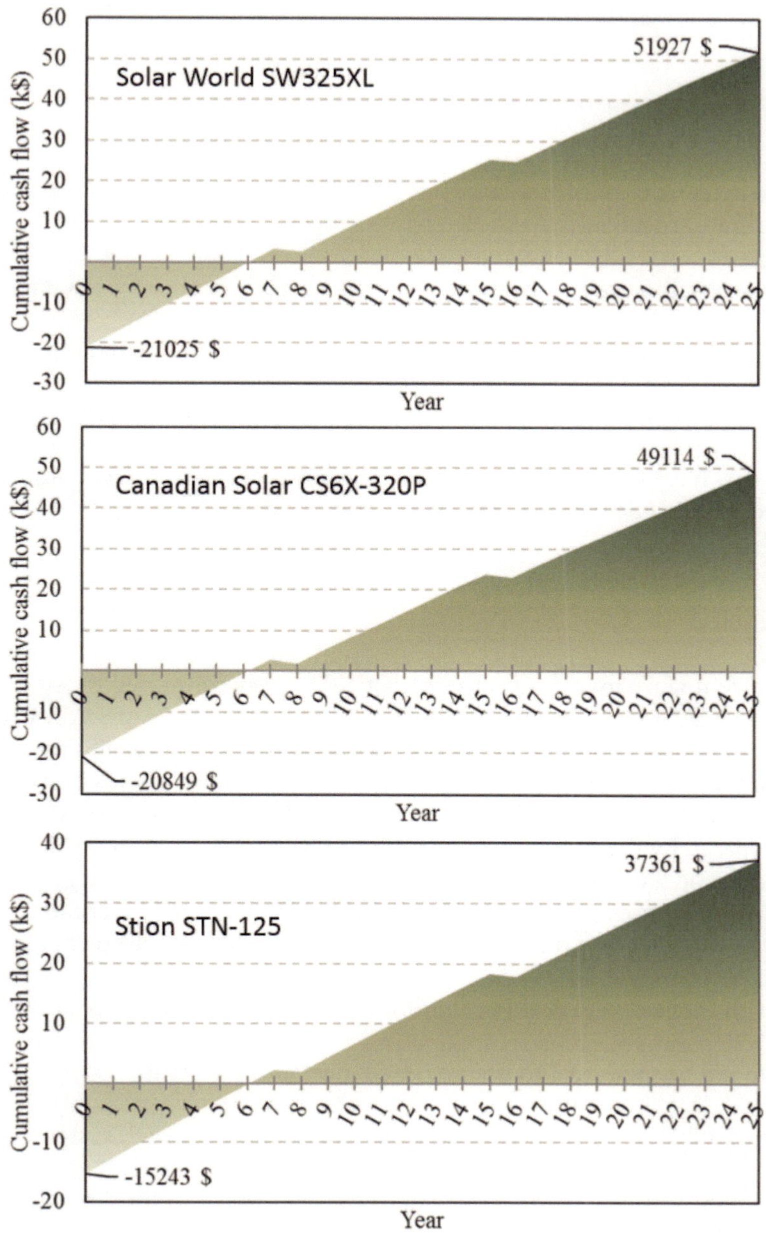

Figure 19.
Economic comparison between different modules of the best economic performance.

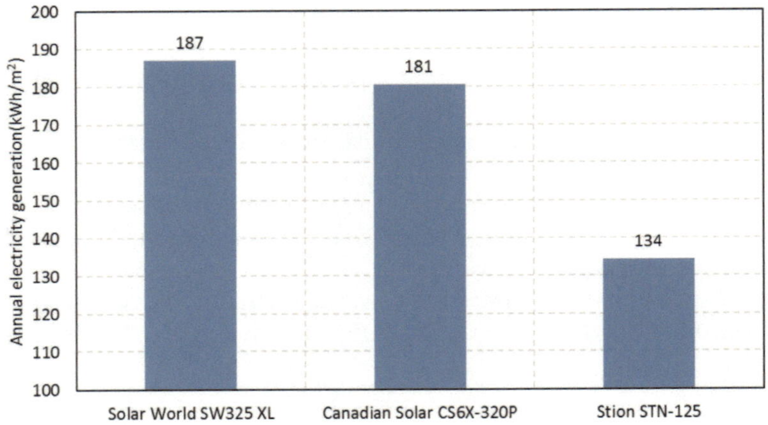

Figure 20.
Energy comparison between different modules of the best economic performance.

the best brands from the economic viewpoint for monocrystalline, polycrystalline, and thin film, respectively. As shown:

- Monocrystalline can give electricity at 3% cheaper than polycrystalline.
- A thin film can give electricity at 2% cheaper than monocrystalline module.
- A thin film can give electricity at 5% cheaper than polycrystalline module.

However, from the energy viewpoint, it seems that the monocrystalline module has the best economic performance (see **Figure 20**).

7. Conclusion

The current work investigates the potential of a grid-connected PV system under the meteorological and working conditions of Padiham (53.5 N, 2.3 W), UK. The obtained results present a comparison between different PV modules types, including thin film, polycrystalline and monocrystalline. Different brands of each type were considered in this work. The brands (**Table 1**) were chosen based on the report prepared by the Fraunhofer Institute for Solar Energy Systems, ISE [11]. An excel-based model was developed to simulate the economic and energy performance of the proposed system. The model was used to define the optimal slope angle of the module. Besides, the best module type and brand from energy and economic performance are specified.

Under the considered working conditions in the UK, simulations show that:

- There is a big potential to use a grid-connected PV system.
- Up to 226 kWh/y can be generated per square meter of such system.
- The initial cost of the system ranged from $15,000 (thin film) to $36,000 (monocrystalline).
- Initial cost per installed capacity is:

- 1.54 $/W thin-film-based system

- 1.84 $/W monocrystalline-based system

- The payback time of the systems varies from 6 to 10 years.

Another important conclusion that can be drawn from the results achieved in this work is that the energy and economic performance of a PV module are contradicting. In other words, for each module type, the brand that shows the best energy performance is not the same brand of the best economic performance. Therefore, the preference of the decision-maker should be considered as an essential factor before choosing the PV type and brand.

Comparing the results from different module types and brands show that:

- Motech XS72D3-320 which is a monocrystalline module shows the best energy performance compared to other monocrystalline brands, while Solar World SW325XL shows the best economic performance.

- Regarding polycrystalline module type, among the considered brands, Hanwha Solar One 310 and Canadian Solar CS6X-320P show the best energy and economic performance, respectively.

- Among the selected thin-film module brands, Stion STO-150 shows the best energy performance, while Stion STN-125 shows the best economic performance.

Finally, simulations indicate that monocrystalline module type shows the best energy performance, while thin-film module type shows the best economic performance at the current conditions in the UK.

Acknowledgements

The authors are grateful to EC DREEAM (H2020-EeB-2015, GA 680511) project that is funded by the EC and supported by Chalmers Area of Advance Energy.

Author details

Mohamad Kharseh* and Holger Wallbaum
Chalmers University of Technology, Gothenburg, Sweden

*Address all correspondence to: mohamad.kharseh@chalmers.se

References

[1] Coyne D. World Oil 2018–2050: World Energy Annual Report (Part 2) [Internet]. 2018. Available from: http://peakoilbarrel.com/world-oil-2018-2050-world-energy-annual-report-part-2/ [Accessed: 07-05-2019]

[2] IEA. Global Energy & CO2 Status Report: The Latest Trends in Energy and Emissions in 2018. International Energy agency 2019. Available from: www.iea.org/geco/renewables/

[3] Jackson R, Le Quéré C, Andrew R, Canadell J, Peters G, Roy J, et al. Warning signs for stabilizing global CO2 emissions. Environmental Research Letters. 2017;**12**:110202

[4] Moomaw W, Yamba F, Kamimoto M, Maurice L, Nyboer J, Urama K, et al. Introduction. In: IPCC Special Report on Renewable Energy Sources and Climate Change Mitigation. Cambridge, United Kingdom and New York, NY, USA: Cambridge University Press; 2011

[5] Bilgili M. Hourly simulation and performance of solar electric-vapor compression refrigeration system. Solar Energy. 2011;**85**:2720-2731. DOI: 10.1016/j.solener.2011.08.013

[6] Wrobel J, Sanabria Walter P, Schmitz G. Performance of a solar assisted air conditioning system at different locations. Solar Energy. 2013; **92**:69-83. DOI: 10.1016/j.solener.2013.02.030

[7] Nyholm E, Goop J, Odenberger M, Johnsson F. Solar photovoltaic-battery systems in Swedish households—Self-consumption and self-sufficiency. Applied Energy. 2016;**183**: 148-159. DOI: 10.1016/j.apenergy.2016.08.172

[8] IRENA. Renewable Capacity Statistics 2019. International Renewable Energy Agency 2019. ISBN 978-92-9260-123-2

[9] Taylor M, Daniel K, Ilas A, So E. Renewable Power Generation Costs in 2014 [Internet]. 2015. Available from: www.irena.org/documentdownloads/publications/irena_re_power_costs_2014_report.pdf [Accessed: 08-06-2016]

[10] Helen EH. Humphries. Evaluation of PV Systems in Gårdsten [thesis]. Göteborg: Chalmers University of Technology; 2013

[11] Fraunhofer I. Photovoltaics Report [Internet]. 2016. Available from: http://ecee.colorado.edu/~ecen5009/Resources/Photovoltaics/Fraunhofer2016.pdf [Accessed: 15-12-2016]

[12] Al-Khawaja MJ. Determination and selecting the optimum thickness of insulation for buildings in hot countries by accounting for solar radiation. Applied Thermal Engineering. 2004;**24**: 2601-2610. DOI: 10.1016/j.applthermaleng.2004.03.019

[13] Kharseh M, Wallbaum H. How adding a battery to a grid-connected photovoltaic system can increase its economic performance: A comparison of different scenarios. Energies. 2019; **12**:19. DOI: 10.3390/en12010030

[14] El Chaar L, Lamont LA. Global solar radiation: Multiple on-site assessments in Abu Dhabi, UAE. Renewable Energy. 2010;**35**:1596-1601. DOI: 10.1016/j.renene.2009.10.007

[15] Basunia MA, Yoshiob H, Abec T. Simulation of solar radiation incident on horizontal and inclined surfaces. Journal of Engineering Research. 2012;**9**:27-35. DOI: 10.24200/tjer.vol9iss2pp27-35

[16] Duffie JA, Beckman WA. Solar engineering of thermal processes. 4th ed. New Jersey: Wiley; 2006. p. 910. DOI: 10.1002/9781118671603

[17] Yoon K, Yun G, Jeon J, Kim KS. Evaluation of hourly solar radiation on inclined surfaces at Seoul by photographical method. Solar Energy. 2014;**100**:203-216. DOI: 10.1016/j.solener.2013.11.011

[18] Tukiainen M. Gaisma: World Sunrise, sunset, dawn and dusk times [Internet]. 2016. Available from: www.gaisma.com/ [Accessed: 10-06-2016]

[19] Al-Sabounchi AM, Yalyali SA, Al-Thani HA. Design and performance evaluation of a photovoltaic grid-connected system in hot weather conditions. Renewable Energy. 2013;**53**: 71-78. DOI: 10.1016/j.renene.2012.10.039

[20] Torres-Ramírez M, Nofuentes G, Silva JP, Silvestre S, Muñoz JV. Study on analytical modelling approaches to the performance of thin film PV modules in sunny inland climates. Energy. 2014;**73**: 731-740. DOI: 10.1016/j.energy.2014.06.077

[21] Dalenbäck J. Personam communication with Jan-Olof Dalenbäck, Professor in Building Services Engineering/Project manager CIT Energy Management AB. Jan-Olof.Dalenback@chalmers.se. 2017

[22] Paudel AM, Sarper H. Economic analysis of a grid-connected commercial photovoltaic system at Colorado State University-Pueblo. Energy. 2013;**52**: 289-296. DOI: 10.1016/j.energy.2013.01.052

[23] Harder E, Gibson JM. The costs and benefits of large-scale solar photovoltaic power production in Abu Dhabi, United Arab Emirates. Renewable Energy. Shenzhen, Guangdong: JA Energy Co; 2011;**36**:789-796. DOI: 10.1016/j.renene.2010.08.006

[24] Khalid A, Junaidi H. Study of economic viability of photovoltaic electric power for Quetta—Pakistan. Renewable Energy. 2013;**50**:253-258. DOI: 10.1016/j.renene.2012.06.040

[25] Chandel M, Agrawal GD, Mathur S, Mathur A. Techno-economic analysis of solar photovoltaic power plant for garment zone of Jaipur city. Case Studies in Thermal Engineering. 2014;**2**: 1-7. DOI: 10.1016/j.csite.2013.10.002

[26] Kaplanis S, Kaplani E. Energy performance and degradation over 20 years performance of BP c-Si PV modules. Simulation Modelling Practice and Theory. 2011;**19**:1201-1211. DOI: 10.1016/j.simpat.2010.07.009

[27] CIA. World Factsbook [Internet]. 2016. Available from: www.cia.gov/library/publications/the-world-factbook/ [Accessed: 15-06-2016]

[28] Skoczek A, Sample T, Dunlop ED. The results of performance measurements of field-aged crystalline silicon photovoltaic modules. Progress in Photovoltaics: Research and Applications. 2009;**17**:227-240. DOI: 10.1002/pip.874

[29] Sánchez-Friera P, Piliougine M, Pelaez J, Carretero J, Sidrach de Cardona M. Analysis of degradation mechanisms of crystalline silicon PV modules after 12 years of operation in southern Europe. Progress in Photovoltaics: Research and Applications. 2011;**19**:658-666. DOI: 10.1002/pip.1083

[30] Flowers ME, Smith MK, Parsekian AW, Boyuk DS, McGrath JK, Yates L. Climate impacts on the cost of solar energy. Energy Policy. 2016;**94**:264-273. DOI: 10.1016/j.enpol.2016.04.018

[31] Eurostat Statistics Explained. Electricity Prices for Household Consumers [Internet]. 2016. Available from: http://ec.europa.eu/eurostat/statistics-explained/index.php/File:Electricity_prices_for_household_consumers_second_half_2014_(%C2%B9)_(EUR_per_kWh)_YB15.png [Accessed: 15-04-2016]

[32] The Global Economy. Sweden: Real Interest Rate [Internet]. 2016. Available from: http://www.theglobaleconomy.com/Sweden/Real_interest_rate/ [Accessed: 10-04-2016]

[33] Tom J, Alice L. Personal communication with JA Energy Co., Limited. Shenzhen, Guangdong: JA Energy Co; 2016

Chapter 3

Analysis of the Solar Tracking System for a Mobile Robot Prototype

Dario Guilherme Toginho, Alex Archela and Leonimer Flávio de Melo

Abstract

Methods to detect the position of the sun and orientate a solar panel to its position are used in order to optimize the power generated. Two possible approaches are using light depending resistor (LDR) sensors, or using a GPS and equations that model the geometry between the Earth and the Sun. The main objective of this work is the proposal of a prototype system to optimize the harvesting of solar energy on photovoltaic panels. In this chapter, a mobile robot powered by a solar panel orientated by a LDR matrix and a GPS device was developed. The LDRs were used as the representation of vectors normal to a surface, where its sum resulted in the most lighted point. The GPS, in turn, provided location, date and time data, which were used in the calculations of the Sun's azimuth and zenith, used to orientate the panel. The obtained results show that an orientated photovoltaic panel has a better performance when compared to a static panel. Possibilities and opportunities of research tend to grow for the next years with many possible works to be done in the future, both in mobile robotics and in other systems powered by photovoltaic panels.

Keywords: photovoltaics, embedded, robotics, tracking, optimization

1. Introduction

The use of photovoltaic energy is one of the most promising alternative energy resources. Given the high fossil fuel consumption and economic and environmental impact of other sources, solar energy has become one of the most viable alternatives, given its low impact and the fact that it is renewable.

The sun provides both thermal and light energies, and in order to convert its light energy to electrical energy, photovoltaic cells are used. The cells of the panel used in this work are made of silicon. According to [1], the efficiency of large-area commercial cells is about 24%, and the upper limit of the silicon solar cell efficiency is 29%. In order to optimize the energy generated by a photovoltaic panel, instead of using it in a fixed position, methods of tracking the position of the sun and adjusting the panel accordingly are used, so the sun's rays are perpendicular to the photosensitive surface [2].

The works of [3, 4] show systems that use light-dependent resistor (LDR) as a way to obtain the sun's position [5]. On the other hand, the work of [6] shows an approach that allows the control of a photovoltaic cell based on its temperature, and the work of [7] shows a different design to the motors that control the position of photovoltaic panels, in order to decrease the energy used by the drivers [8].

In this chapter a method that uses both GPS and a LDR matrix is described, with the solar panel attached to a mobile robot, using ARM microcontroller device (STM32F407). Also there is a comparison of both approaches, in order to find the most efficient position for the panel and the possible applications both in the field of mobile robots and in others.

2. Solar tracking approaches

Similar to the work of [9], in this paper a solar panel powers a mobile robot, but the main difference is that the panel on the robot of [9] has two motors to move it in the x-axis and y-axis [10]. The robot used in this work has a panel that moves in the φ-axis (spherical coordinates), and the adjustment of the panel position consists in operating this φ-axis motor and adjusting the position of the robot [11, 12].

2.1 LDR approach

The LDR matrix consists of a Styrofoam half-sphere with 20 LDRs attached on its surface, each one representing a coordinate of the three-dimensional space. **Figure 1** shows a 3D model representing the LDR positions.

The LDRs have a voltage divider, and the microcontroller's A/D converter reads its value. As light reaches the LDR, its resistance decreases. The design of the

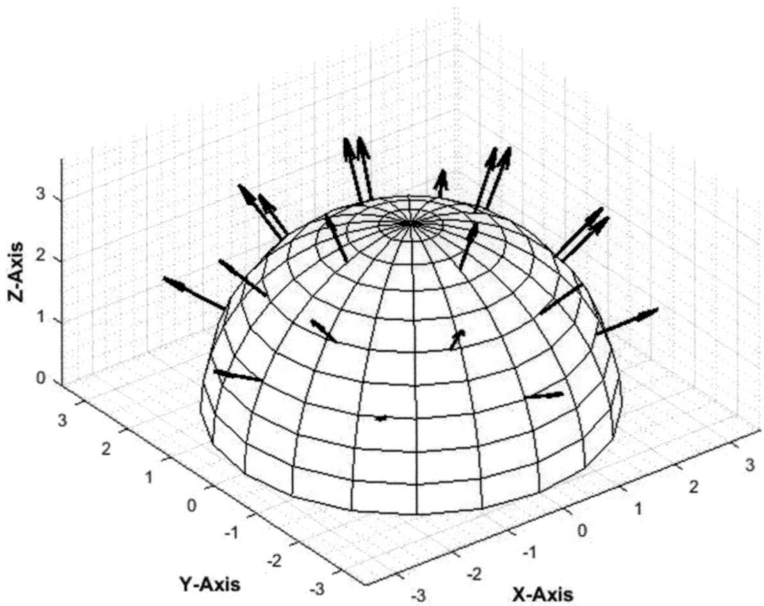

Figure 1.
LDR surface 3D model, where the black arrows represent the position of the LDRs.

voltage divider was done in a way that its output voltage increases, so the more light reaches on the LDR, the higher the readings will be on the A/D converter.

However, the LDRs are sensitive not just to sunlight but also the environment's reflected lights, and the A/D converter reads these lights with a small difference between them. In order to find which reading corresponds to the direct sunlight, a series of mathematical operations are performed, in a way that increases this difference.

When representing the LDRs in spherical coordinates (ρ, θ, φ), we may change ρ to the A/D converter read value. So, we can write the coordinates as $(L(\theta, \varphi), \theta, \varphi)$, where $L(\theta, \varphi)$ is the LDR A/D reading on the corresponding θ and φ coordinates. Since the microcontroller A/D converter has a 12 bit resolution, its value range is seen on Eq. (1):

$$0 \leq L(\theta, \varphi) \leq 4095 \tag{1}$$

In order to find the most relevant readings, all of them are divided by the average and then squared. In this way, it is possible to obtain the most significant readings. These readings are represented by $W(\theta, \varphi)$, as seen on Eq. (2), where $L(\acute{\theta}, \varphi)$ represents the average of the readings of all LDRs:

$$W(\theta, \varphi) = \left(\frac{L(\theta, \varphi)}{L(\acute{\theta}, \varphi)} \right) \tag{2}$$

The next step to obtain the position of the sun via the LDRs is to convert the spherical coordinates $(W(\theta, \varphi), \theta, \varphi)$ to the Cartesian coordinate system (x, y, z) and then sum all elements. As a result, we obtain the coordinates of the position on the Styrofoam surface that represents a vector pointing towards the sunlight. The spherical coordinates to Cartesian equations can be seen on Eqs. (3)–(5):

$$x = W(\theta, \varphi)\, sin\,(\varphi)\, cos\,(\theta) \tag{3}$$

$$y = W(\theta, \varphi)\, sin\,(\varphi)\, sin\,(\theta) \tag{4}$$

$$z = W(\theta, \varphi)\, cos\,(\varphi) \tag{5}$$

Then, the result in spherical coordinates $\left(\rho_{final}, \theta_{final}, \varphi_{final}\right)$ is given by Eqs. (6)–(8), where Σx, Σy, and Σz are the sum of the respective Cartesian coordinates of all LDRs:

$$\rho_{final} = \sqrt{(\Sigma x)^2 + (\Sigma y)^2 + (\Sigma z)^2} \tag{6}$$

$$\theta_{final} = cos^{-1}\left(\frac{\Sigma y}{\Sigma x}\right) \tag{7}$$

$$\varphi_{final} = sin^{-1}\left(\frac{\Sigma z}{\sqrt{(\Sigma x)^2 + (\Sigma y)^2 + (\Sigma z)^2}}\right) \tag{8}$$

Having the θ_{final} and φ_{final} coordinates, it is possible to adjust the position of the panel in such way it is perpendicular do the direct sunlight, using θ_{final} to adjust the position of the robot and φ_{final} to adjust the servomotor.

2.2 GPS approach

The use of the GPS device to obtain the position of the sun is based on a set of equations that return the coordinates of the sun in the horizontal coordinate system. These coordinates are the zenith angle, the altitude angle, and the azimuth angle, and all of them are seen in **Figure 2**.

In this work, Eqs. (9)–(11) are the same from [9]:

$$\theta_z = cos^{-1}[sin(\delta) sin(\varphi) + cos(\delta) cos(\varphi) cos(\omega)] \tag{9}$$

$$\alpha = 90^o - \theta_{Zn} \tag{10}$$

$$A = cos^{-1}\left[\frac{sin(\alpha) sin(\varphi) - sin(\delta)}{cos(\alpha) cos(\varphi)}\right] \tag{11}$$

where θ_z: incidence angle (or zenith angle); δ: declination angle; α: altitude angle; φ: local latitude; ω: hour angle; A: Azimuth angle.

The declination angle δ is the angle between the equator and a line connecting the earth and the sun, and it varies according to the current date. The hour angle ω is the angle that depends on the current time, representing the position of the sun above the observer.

The GPS device used is shown in **Figure 3**, and it provides the current time and location and, with these data, is possible to calculate the solar angles and correct the panel position accordingly.

It sends data on an interval of 1 second. The data received is a string that uses National Marine Electronics Association (NMEA) specification. The string contains coded phrases, and each phrase contains a specific set of information. Among the phrases, there is information containing the current location, time, and date. **Figure 4** shows an example of all phrases received.

The first phrase, being the one with the code $GPRMC, contains all the information required to obtain the current position of the sun. Commas separate each part of the data. After the code, the first number is 134759.00, which represents the time 13h47min59s. The second number is 2319.68361 followed by the letter S, which

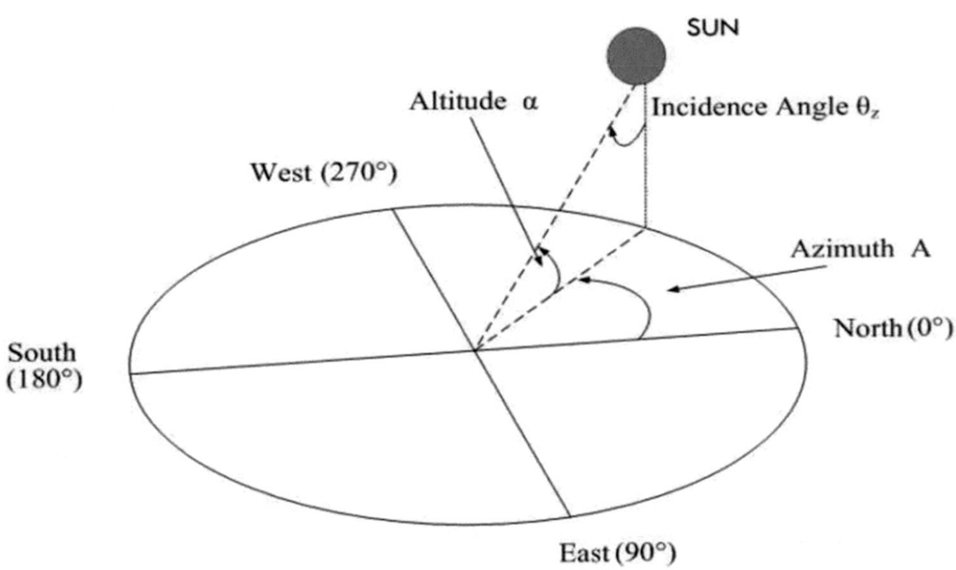

Figure 2.
Azimuth and altitude angle of the sun (adapted from [6]).

Figure 3.
Picture of GPS device used.

```
$GPRMC,134759.00,A,2319.68361,S,05112.15109,W,0.019,,170518,,,A*78
$GPVTG,,T,,M,0.019,N,0.034,K,A*2C
$GPGGA,134759.00,2319.68361,S,05112.15109,W,1,05,2.14,602.5,M,-1.5,M,,*4C
$GPGSA,A,3,21,11,31,08,18,,,,,,,,3.44,2.14,2.70*03
$GPGSV,4,1,13,01,03,232,12,03,02,285,,08,46,236,28,11,13,223,33*74
$GPGSV,4,2,13,18,23,221,29,20,01,115,,21,14,070,36,22,19,272,25*75
$GPGSV,4,3,13,26,02,357,,27,69,309,26,31,13,029,38,32,,,27*4B
$GPGSV,4,4,13,33,42,061,34*4D
$GPGLL,2319.68361,S,05112.15109,W,134759.00,A,A*63
```

Figure 4.
NMEA specification data received through serial monitor.

represents south latitude 23°1,968,361′. After that is the number representing the longitude followed by the letter W, which stands for west, so the longitude is 51° 1215109'West. The next piece of information, which is 0.019, is the current speed over the surface in knots, followed by the number 170518, which stands for the date, in this case meaning May 17th of 2018.

3. Implementation

A set of PWM signals drive the motors of the robot, and its speed is proportional to the duty cycle used. The duty cycle in turn is proportional to the difference between the direction of the robot and the direction of the sun. A picture of the robot is seen in **Figure 5**.

Figure 5.
Picture of the mobile robot.

This difference is obtained calculating the Euclidian distance between the obtained position of the sun and the direction of either the front or the back of the robot (whichever is the nearest one) and minimizing it. As this difference is calculated using unit vectors, its range, as seen on Eq. (12), is never higher than 2:

$$0 \leq d \leq 2 \tag{12}$$

So d is the maximum possible distance between the position of the robot and the position of the sun; therefore, $d = 2$ implies a maximum use of the motors to correct the position of the robot, and $d = 0$ means the robot is at the desired position.

However, as a control method, $d = 0$ is not used as a parameter to determine if the desired position is achieved, for it is an ideal situation. Instead, a value close to 0 is used. So, for example, a possible condition to consider that the robot is in the correct position is to consider $d < 0.01$.

Sections 3.1 and 3.2 have a detailed description of each approach used to detect the direction of the sun, and **Figure 6** contains a block diagram of the system.

3.1 LDR implementation

A half-sphere of Styrofoam was used to arrange the LDRs symmetrically, and the position of each LDR was chosen based on the manufacturing marks on its surface, having a total of 20 positions. All LDRs had a common terminal connected to a voltage source, and the other ends were connected to a multiplexer circuit. This multiplexer circuit is connected to the microcontroller, which selects the LDR and also reads its analogic value.

After the microcontroller calculates the position of the sun, it sends a signal to the motors of the mobile robot, to adjust its position in a way that its front or its back faces the sun and to the servomotor to adjust it to the proper inclination. The decision of if it is either the front or the back of the robot that should face the sun is based on which of these sides are closer to sunlight, in order to minimize the motion of the robot.

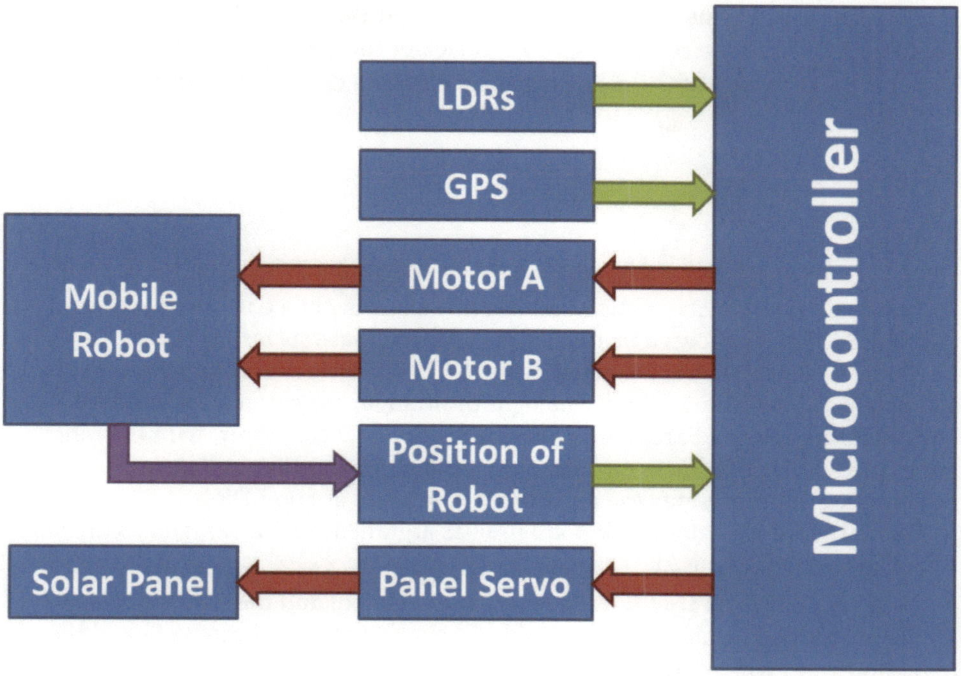

Figure 6.
Block diagram of the system.

The signal sent to the motors of the robot, as said in the beginning of Section 3, is a PWM signal, proportional to the Euclidian distance in the x-y plane of the position of the sun and the front, or back, of the robot, which is seen in Eq. (13):

$$d_{ldr} = \sqrt{cos^2(\theta_{final}) + \left[sin\,(\theta_{final}) - y_{robot}\right]^2} \tag{13}$$

The value of y_{robot} is either 1, if the sun is closer to the front of the robot, or -1, if the sun is closer to the back of the robot. As d_{ldr} gets closer to 0, it means the robot is getting closer to the direction of the sun.

3.2 GPS implementation

An electronic GPS device was used to obtain the required data to calculate the position of the sun, these data being latitude, longitude, day, month, year, hour, and minute. In addition, an electronic compass was used to obtain the angle the front of the robot formed with the north direction.

In order to keep the same reference as the LDR approach, the compass returns 90° when the front of the robot faces north and −90° when its back faces north. The value of the zenith angle is used as a signal in the servomotor to adjust the inclination of the solar panel. To make the robot face the sun, the Euclidian distance between its front or its back, depending on which one is closer, and the azimuth of the sun must be minimized. Eq. (14) returns this distance, where θ_n is the angle of the north direction and the robot:

$$d_{gps} = \sqrt{cos^2(\theta_n - A) + \left[sin\,(\theta_n - A) - y_{robot}\right]^2} \tag{14}$$

As in Section 3.1, the value of y_{robot} is either 1, if the sun is closer to the front of the robot, or -1, if the sun is closer to the back of the robot. As d_{gps} gets closer to 0, it means the direction of the robot is getting closer to the direction of the sun.

4. Results

4.1 LDR results

The preliminary tests on the LDR matrix were done in a darkroom with a flashlight. **Figure 7** shows a picture of the position that was tested.

Then, with each LDR connected to the A/D converter, the data were collected and processed in order to obtain the light position on the surface. In **Table 1**, written in the color red are the values read by the A/D converter, with the range mentioned on Eq. (1), for each LDR on its respective position in **Figure 7**, and NA (not applicable) represents the positions where there are no LDRs.

Then, the values from **Table 1** are used as argument to Eq. (2), and as the next step, Eqs. (3)–(8) are used.

Using the data of **Table 1**, the average is calculated, and the value $L(\acute{\theta}, \varphi) =$ 1789.5 is obtained. So, all values in **Table 1** below 1789.5 will become 0, and the remaining ones will be subtracted and then squared.

The sum of these values as vectors equals the resulting position $(\rho_{final}\theta_{final}\varphi_{final})$, which is seen as the blue arrow in **Figure 8**. The spherical coordinates obtained are (24.952508°, 38.555257°, 97.442222°). However, in **Figure 8**, the ρ coordinate was reduced in order to fit in the figure.

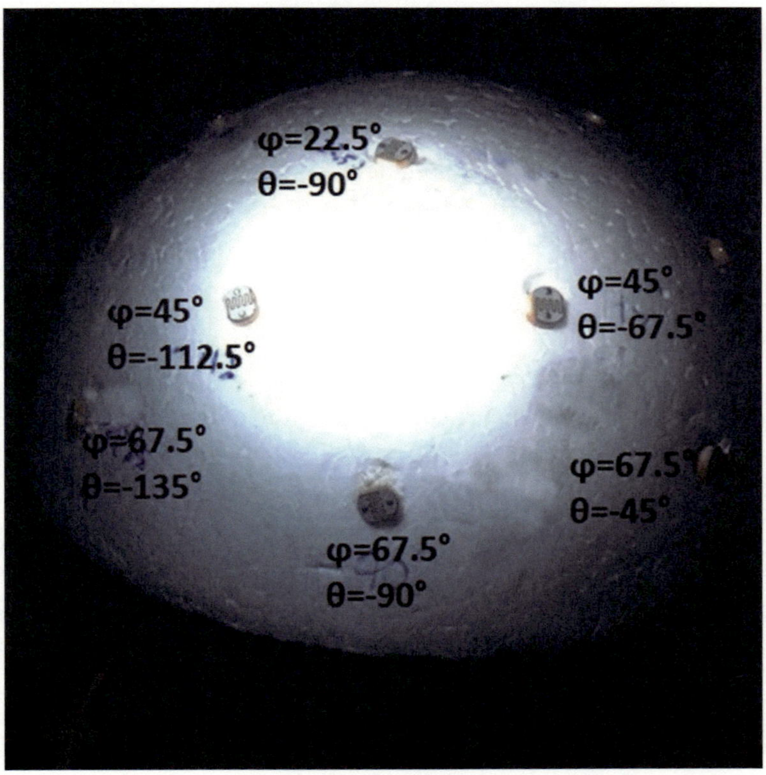

Figure 7.
Photo of the first LDR sample with the coordinate of each LDR.

	$L(\theta, \varphi)$		
LDR spherical coordinates	$\varphi = 22.5$	$\varphi = 45$	$\varphi = 67.5$
$\theta = 0°$	2580	NA	840
$\theta = 22.5°$	NA	481	NA
$\theta = 45°$	NA	NA	429
$\theta = 67.5°$	NA	361	NA
$\theta = 90°$	914	NA	281
$\theta = 112.5°$	NA	292	NA
$\theta = 135°$	NA	NA	154
$\theta = 157.5°$	NA	1673	NA
$\theta = 180°$	2647	NA	1151
$\theta = 202.5°$	NA	3080	NA
$\theta = 225°$	NA	NA	2478
$\theta = 247.5°$	NA	3674	NA
$\theta = 270°$	3616	NA	3433
$\theta = 292.5°$	NA	3668	NA
$\theta = 315°$	NA	NA	2371
$\theta = 337.5°$	NA	1667	NA

Table 1.
Raw readings of the a/D converters with range from 0 to 4095.

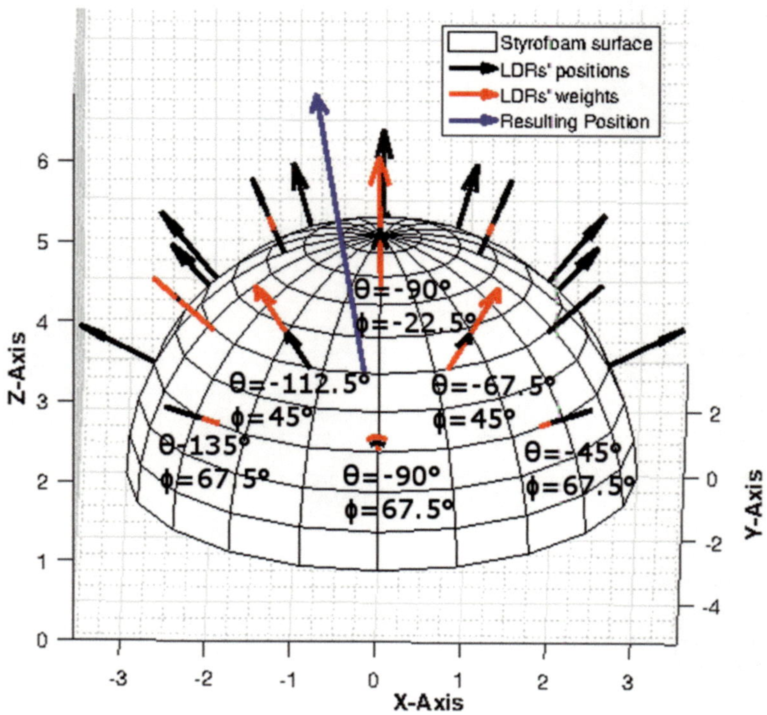

Figure 8.
3D model of the collected data and its result on the surface.

Table 2 shows a set of 10 samples of **Figure 7** on a 1 second interval, in which $n = 1$ is the sample plotted in **Figure 8**. The table also includes the mean value of the samples and the standard deviation.

4.2 GPS results

The GPS was tested under the following conditions:

- Local: Londrina State University
- Time: 13h48min
- Sample time: 1 s

The GPS device received its fix data in the default interval of 1 second, and for each sample, an altitude angle and an azimuth angle were calculated. **Table 3** shows the solar angle data and its mean values and standard deviations.

4.3 Power results

Three test scenarios were executed. One is where the solar panel is fixed and the other two where the sun is tracked, one using LDR and the other using the GPS. The sampling period was 5 minutes.

In **Figure 9** it is possible to observe the power generated by the panel on each scenario. The LDR and the GPS approaches have shown a very similar performance compared to each other, and both of them demonstrated to be much better than a fixed panel, having a gain of approximately 30%. The yellow line in the figure represents what would be the ideal maximum generated power by the panel,

Sample time (seconds)	φ coordinate [°]	θ coordinate [°]
1	38.555257	−97.442222
2	38.422000	−99.008884
3	38.570818	−97.637040
4	38.551388	−97.037137
5	38.575064	−98.695924
6	38.503982	−98.473986
7	38.545190	−100.300662
8	38.470601	−96.550539
9	38.586206	−97.503065
10	38.445780	−98.427292
Mean φ [°]	38.524916	
Std. deviation of φ [°]	0.068990	
Mean θ [°]	−97.781321	
Std. deviation of θ [°]	0.855672	

Table 2.
LDR samples over 10 s.

Sample time (seconds)	Altitude α [°]	Azimuth A [°]
1	42.092760	−28.719862
2	42.095466	−28.704564
3	42.098172	−28.689464
4	42.100877	−28.674137
5	42.101605	−28.714010
6	42.090841	−28.697598
7	42.096462	−28.700248
8	42.095976	−28.684667
9	42.097934	−28.675514
10	42.097912	−28.675200
Mean α **[°]**	42.096819	
std. deviation of α **[°]**	0.003493	
Mean A **[°]**	−28.697007	
std. deviation of A **[°]**	0.019659	

Table 3.
GPS samples over 10 seconds.

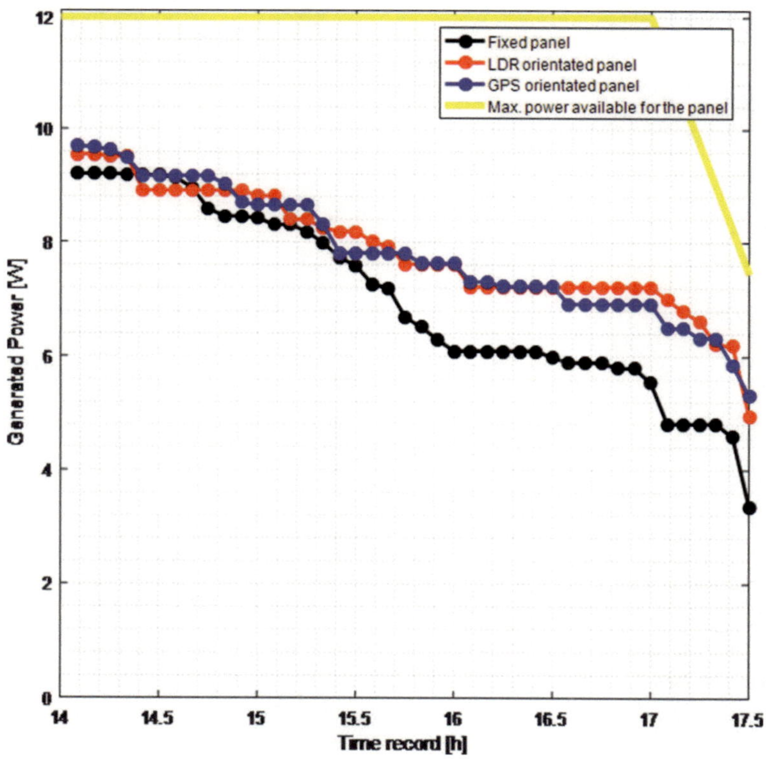

Figure 9.
Power samples of the three tested scenarios.

considering the average solar irradiation nearby the site and the surface area of the panel. The LDR approach as the GPS has shown an efficiency of 80% compared to the ideal maximum power.

5. Conclusion

During the data collection, the LDR matrix has showed to be very sensitive to noise, so it presented a higher deviation. On the other hand, the GPS data had a much lower deviation. However, during the data collection, it took a certain amount of time to get a fix and obtain the current date, time, and location.

Besides the solar panel orientation, i.e., for applications in the mobile robotics, the GPS device provides navigation data that may be useful, which is an advantage. However, if the device takes too long to get a fix or loses the fix, the data will be null or delayed. So, in order for a mobile robot to obtain full-time orientation, both the GPS and LDR approaches may be used together.

For other applications, such as a panel on a fixed position, the latitude and longitude are fixed, so there is no need for a GPS device, just a real-time clock (RTC) device, which eliminates the GPS fix time issue and also has a lower cost compared to the LDR approach.

Having all things considered, the LDR is a low-cost method to track the position of the sun; however, it is more affected by noise. The sun-earth equations have shown problems only when depending of a GPS device to provide data; otherwise, it is the most precise approach.

Author details

Dario Guilherme Toginho, Alex Archela and Leonimer Flávio de Melo*
State University of Londrina, Londrina, PR, Brazil

*Address all correspondence to: leonimer@uel.br

References

[1] Blakers A, Zin N, McIntosh KR, Fong K. High efficiency silicon solar cells. Energy Procedia. 2013;**33**:1-10

[2] Fedor P, Perduková D. Use of fuzzy logic for design and control of nonlinear MIMO systems. In: Ramakrishnan S, editor. Modern Fuzzy Control Systems and Its Applications. London, UK: IntechOpen; 2017. DOI: 10.5772/68050

[3] Mustafa FI, Al-Ammri AS, Ahmad FF. Direct and indirect sensing two-axis solar tracking system. In: 8th International Renewable Energy Congress (IREC). 2017. pp. 1-4

[4] Mustafa FI, Shakir S, Mustafa FF, Naiyf AT. Simple design and implementation of solar tracking system two axis with four sensors for Baghdad city. In: 9th International Renewable Energy Congress (IREC). 2018. pp. 1-5

[5] Ikedi C. Experimental study of current-voltage characteristics for fixed and solar tracking photovoltaics systems. In: Prabaharan N, Rosen MA, Campana PE, editors. Recent Developments in Photovoltaic Materials and Devices. London, UK: IntechOpen; 2019. DOI: 10.5772/intechopen.79710

[6] da Rocha NMM, Passos JC. MPPT method based on temperature control of the photovoltaic cells. In: 12th IEEE International Conference on Industry Applications (INDUSCON). 2016. pp. 1-8

[7] Kuhn VN, Gonzatti F, Franchi D, Miotto M, Camargo MN, Farret FA. Hybrid motor driver for solar tracking system. In: 12th IEEE International Conference on Industry Applications (INDUSCON). 2016. pp. 1-6

[8] Chin CS. Model-based simulation of an intelligent microprocessor-based standalone solar tracking system. In: Katsikis VN, editor. MATLAB–A Fundamental Tool for Scientific Computing and Engineering Applications. Vol. 3. London, UK: IntechOpen; 2012. DOI: 10.5772/46458

[9] Fei Y, Lv H. Design of the solar-driven module on modular mobile robot. In: 19th International Conference on Mechatronics and Machine Vision in Practice (M2VIP). 2012. pp. 470-473

[10] Milea L, Zafiu A, Oltu O, Dascalu M. Theory, algorithms and applications for solar panel MPP tracking. In: Manyala R, editor. Solar Collectors and Panels, Theory and Applications. London, UK: IntechOpen; 2010. DOI: 10.5772/10337

[11] Zhan TS, Lin WM, Tsai MH, Wang GS. Design and implementation of the dual-axis solar tracking system. In: IEEE 37th Annual Computer Software and Applications Conference. 2013. pp. 276-277

[12] Engin D, Engin M. Sensors and digital signal conditioning in mechatronic systems. In: Yildirim S, editor. Design, Control and Applications of Mechatronic Systems in Engineering. London, UK: IntechOpen; 2017. DOI: 10.5772/67986

Section 2

Lifetime and Reliability Issues

Chapter 4

Some Reliability Aspects of Photovoltaic Modules

Titu-Marius I. Băjenescu

Abstract

Solar cells and photovoltaic modules are energy conversion components that produce electricity when exposed to light. The originality of photovoltaic energy as we understand it here is to directly transform light into electricity. Thin-film silicon in particular is better at low and diffuse illuminations and decreases less than the crystalline when the temperature increases while reducing the amount of material and manufacturing costs. However, the quality of the material and the efficiency of the conversion limit their use on a large scale. If the light absorption of the ultra-thin layers of the active material could be improved, this would lead to low recombination currents, higher open-circuit voltages and higher conversion efficiency. PV systems often communicate with utilities, aggregators and other grid operators over the public Internet, so the power system attack surface has significantly expanded. Solar energy systems are equipped with a range of grid-support functions, which—if controlled or programmed improperly—present a risk of power system disturbances.

Keywords: climatic stress factors, useful life, PV module failure, LCoE, EVA, PID, LID, defects, MWT, field failures, failure analysis, influence of temperature, encapsulants, perspectives, conclusions

1. Introduction

Discovered in 1839 by Antoine Becquerel [1], the photovoltaic effect allows the transformation of light energy into electricity. This principle is based on semiconductor technology. It consists in using photons to release electrons and create a potential difference between the cell's terminals that generate a direct electrical current. All photovoltaic devices have a p-n junction in a semiconductor through which PV voltage develops. Production costs have been reduced by about 20% [2] each time the production volume has doubled. The properties of copper oxide for rectifying alternating current were discovered for some 100 years.

Gravure printing electricity appeared in 1930 with copper oxide cells and then selenium. However, it was not until 1954, with the construction of the first silicon photovoltaic cells in the Bell Telephone company laboratories, that we saw the possibility of supplying energy. Very quickly they were used to power space vehicles in the 1960s with space satellite equipment. Then, from 1970, the first land uses the electrification of isolated sites. During the 1980s, terrestrial photovoltaic technology made steady progress with the installation of several power plants of a few megawatts and even became familiar to consumers through many low-power products using it: watches, calculators, radio and weather beacons, solar pumps and refrigerators. In the early 1990s, when solar developers announced the price per kilowatt hour (kWh) of photovoltaic energy, everyone laughed: we were ten times above the price of electricity purchased from classical generators. Fifteen years later, the costs have been cut by two or three, and no one is laughing anymore. Philippe Malbranche, a specialist in solar technologies at the Atomic Energy Commission (AEC), made it clear: "In the long run, solar energy will prevail. Not because of ideological choice or concern for the environment, but because it will be the most economically profitable."

When we look at the energy that comes directly from the sun per m^2, it is 1000–2000 times more, i.e. 1–2 MWh per square meter per year! About 70% of this energy can be recovered in the form of heat thanks to solar thermal energy: this represents a resource of almost 1 $MWh/m^2/year$. In the case of photovoltaics, with conversion efficiencies of 10–20%, 100–200 $kWh/m^2/year$ can be recovered directly in electrical form.

When a photon interacts with an electron involved in chemical bonds between atoms, it moves it from its equilibrium level (occupied level of the valence band) to a higher excited level of energy (unoccupied level of the conduction band). The electron then returns to equilibrium at its initial level by re-emitting the absorbed energy, which can be in the form of heat, light and chemical energy—as in the case of photosynthesis or electrical energy if it can be recovered. In the latter case, we are talking about photovoltaic energy. In the case of a p-n junction of a semiconductor, an electron–hole pair is created under the action of a photon, the hole being the equivalent of a positive charge. Under the action of the electric field this created at the n-p interface, there is separation of charges and creation of a photocurrent.

At the beginning of studies on the chemistry of plasma-assisted deposition, the growth of deposition was atom by atom. One of the major discoveries was to increase the growth rate of the deposit by generating small clusters in the vapor phase before depositing them on the surface. In this case, silicon nanocrystals are obtained within an amorphous matrix. This technology is currently in full expansion and corresponds to significant yield increase prospects. Although the deposition of each layer requires the control of delicate chemical problems, we now know how to produce and market this type of product on a large scale.

1.1 New thin-film dies: the cadmium telluride dies

The interest in this technology—and this explains its success—is that the deposition processes are extremely fast (from a few seconds to a few minutes), which makes it possible to achieve high production rates and thus reduces production costs (less than one dollar per watt). The arrival of this technology marked a real breakthrough in the competitiveness of photovoltaic.

The copper indium diselenide die (CIS die) is expected to develop significantly as cells based on the same thin-film system achieve yields of more than 20% in the laboratory. Industrial production modules have yields of up to 13%, but this nascent production (1.7% of the market) is in full development. These cells use polycrystalline materials, filled with defects, and grain joints, materials that photovoltaic specialists would not have paid the slightest interest a few years ago. However, through the miracle of understanding the chemistry of these materials and their very complex interfaces, it works and even works very well!

The modulation of thin-film technologies is not based on the serialization of "wafers" glued next to each other, as is the case in conventional silicon technologies. In the case of thin-film modules, all this is done directly on the panel by insulating thin strips, oft using lasers, which are then connected in series with each other.

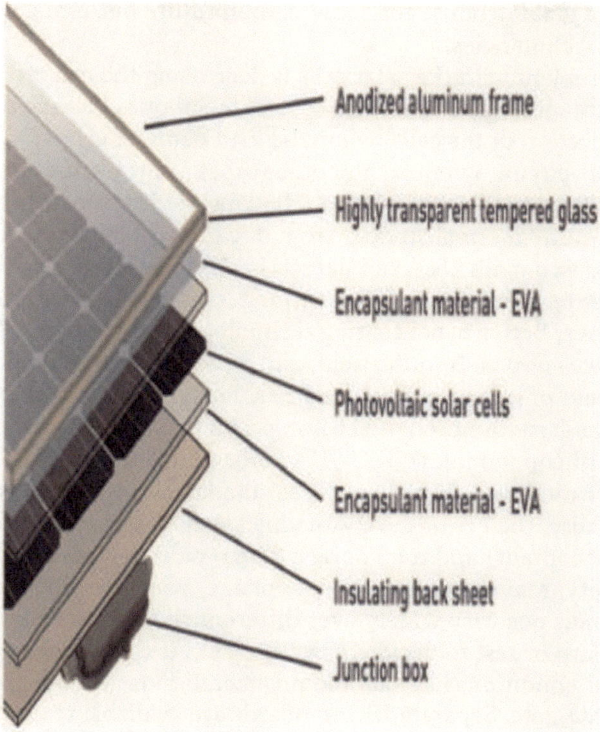

Figure 1.
The structure of a PV module.

This is called the "monolithic" connection. The serial connection allows increasing the voltage delivered by the module by adding the elementary voltages produced by each cell (here by each band). This specificity in the production technique of thin-film cells allow a great flexibility in the models of cells produced and a great adaptability to the needs according to the field of application.

The first amorphous silicon wafers were made in 1980 by Wolf [3] by simulating the ideal parameters for record efficiencies. Reducing the thickness of the layer, the voltage of the open circuit is increased by saturation of the current, as a result of the reduction of the geometric factor. The first publications concerning the manufacture of amorphous silicon cells (a-Si) appeared late, in 1960 [4, 5]. These entities still consist of silicon but only to a thickness of about 1 μm. The first amorphous silicon solar cell was made by Carlson in 1976 [6], and the first products appeared on the market in 1981. The type of system used influences the cost of the system, the stand-alone systems with batteries and auxiliary power generators which are more expensive than grid-connected systems. The interconnected and encapsulated solar cells form a photovoltaic module (**Figure 1**). Solar products have a great promise for a low-carbon future but remain expensive relative to other technologies.

The components of a photovoltaic power system are various types of photovoltaic/solar cells interconnected and encapsulated to form a photovoltaic module[1] (**Figure 1**).

---

[1] Photovoltaic modules are typically rated between 50 W and 300 W with specialized products for building-integrated PV systems at even larger sizes. Quality PV modules are typically guaranteed for up to 25 years by manufacturers.

Solar PV has a great promise for a low-carbon future but remain expensive relative to other technologies.

The ugliest problem with the solar cells is decreasing the cell efficiency with the panel temperature increase. The circuit control techniques include things like (1) using a smaller portion of the cells, when the load demand is low; (2) the wear leveling; (3) short-circuit, overload and other types of protections, current limiting, etc. The solar panels are subjected to the following environmental effects: (a) wind stresses and (b) cyclic thermal stresses from day to night and from summer to winter. These are two main observed defects caused by the repetitive environmental effects. Newer techniques involving varying levels of electroluminescence and photoluminescence, better panel coating technologies preventing debris accumulation on the surfaces of panels in the field, and of course, the technological approaches as ways of mitigating cost and risk currently exist in the industry.

Three IEC standards are used: IEC 61215 part 2 through part 4 (for mono- and multicrystalline silicon modules), IEC 61646 (for thin-film modules) and IEC 62108 (for concentrator modules). However, these standards can be viewed as general guidelines. In reality, the life of the PV modules mainly depends on intensity, distribution (UV content) and total dosage of sun radiation energy, ambient temperature, humidity, maximum operational voltage, wind speed and direction of the site. As these conditions vary widely over different sites, it may not be possible to define one standard or test to decide their lifespan. Further, the operation life depends on other conditions like salt and mist conditions and the presence of gases like NH_3, SO_2, etc. Separate IEC standards are available to address these conditions. Therefore, it is very important to know the site conditions. The components that can fail and be repaired are module failure, string failure, DC combiner failure, inverter failure, transformer failure, AC disconnect failure and tracker failure.

Solar PV power plants are composed of thousands of solar modules. It is a known fact that 2% of them will fail after year 10 of operation, causing losses as high as 27% of total income.

Reduced cost of capital has resulted in the out years having real value in discounted cash flow analysis. The advantages and limitations of PV solar modules for energy generation are reviewed with their reliability limits. The high reliabilities associated with PV modules are indirectly reflected in the output power warranties usually provided in this industry (40-year module lifetimes may not sound as exciting as new photovoltaic materials, but it's essential to make solar power economic). Reliability evaluation based on degradation models is commonly applied in highly reliable products as a cost-effective and confident way of evaluating their reliability. About 80% of the current production uses wafer-based crystalline silicon technology [4, 7].

The development of photovoltaic materials comprised of non-toxic, abundant elements is an important step toward increasing the economic viability of solar energy to meet growing global energy needs.

The performance of a solar cell is a sensitive function of the microstructure of the component materials. Recombination of photo-excited carriers at defects is one of the main contributors to low efficiency. The focus of this contribution is on PV modules with crystalline silicon (c-Si) cells, which represent the dominant technology with over 90% of the market share and of cumulative installations. It is expected that a broad variety of technologies will continue to characterize the PV technology portfolio, depending on the specific requirements and economics of various applications [8].

Among the renewables, the conversion of sunlight into electricity by photovoltaic (PV) devices is a reliable choice to cope with the growing energy consumption.

1.2 Testing problem and interactions caused by adhesion between components

Can a module durability of 30 years be guaranteed by combining components found to have no change in physical properties after 5000 h of pressure cooking testing or high-temperature/high-humidity testing? It seems this approach has been investigated, but there is currently deeply rooted resistance to it; it is difficult to confirm the effect that the gases and oxides generated from encapsulant components have on a variety of other components and difficult to estimate how vapor permeability changes over time when non-uniform tensile and other forces are applied by module creation. It is also impossible to determine the interactions caused by adhesion between components. As a wide variety of components are developed, the development of new reliability test methods (service life verification methods) should inevitably follow in the future [9].

The introduction of a new technology must be competitive with the technologies that it is replacing. It must be cost-effective and ensure an appropriate return on investment (ROI) for investors. The return on investment for governments will take the form of job creation and positive environmental impacts. Each new technology must be capable of generating such a high quantity of revenue that the production costs are reduced to a minimum (economies of scale). In an effort to introduce the use of renewable energy sources, governments worldwide have created incentives to encourage the development of renewable technologies in order to improve our environment. These encouragements help to ensure that the desired volumes are achieved at the early stages of market introduction, so that economies of scale drive the prices to the point of being competitive without the need for other motivations.

PV modules are often considered to be the most reliable component of a photovoltaic system. The alleged reliability has led to the long warranty period for modules up to 25 years. Their real lifespan is unknown; some can last much longer, probably 40 years with very little degradation of their performances. However, since the solar sector is relatively young, it is difficult to have figures in the long term; their conditions and lifespan depend on a wide variety of factors related to their location: a panel installed in Germany will not face the same problems as another used in the Sahara or on the top of a mountain. The panels have a guarantee of a maximum yield loss of 0.7% per year, i.e. after 25 years a maximum loss of about 15%.

When assessing the reliability of a PV system, it is important to consider not only the PV module but the system as a whole. An installed PV system can only provide the expected service if all its components function properly and the complete PV system is properly serviced, from the solar cell to the high-voltage connection.

While it is difficult to measure the performance of a single module in a PV system due to the lack of feedback on the different degradation modes of PV modules, errors that lead to module degradation are generally not taken into account. It is important to remember that, from an economic point of view, consumers are increasingly interested in the reliability and lifetime of their photovoltaic systems. The lifetime and reliability of a PV system largely depends on the energy efficiency of the modules and their different degradation modes. Therefore, research must more and more focus on the degradation of photovoltaic modules.

The panels are materials intended for use under blazing sunlight, in the open air, and therefore subject to the constraints of weather, climate and many other factors. Damage to panels in operation is numerous and can affect different parts of the panels. UV exposure alters the polymers used to encapsulate the photovoltaic cells and modifies the mechanical characteristics of these polymers. Moisture can penetrate the panel and degrade welds, cables and connections. Freezing can cause mechanical stress, while in the desert, sand can erode materials. In very rare cases, the panels can overheat and ignite [10].

While the reliability, quality and lifetime prediction are well established for most products, a complete understanding of these disciplines for PV modules is not yet possible because the desired lifetime is several decades and a first-hand assessment of lifetime predictions is not possible.

However, even today, the risks associated with module in-service performance over long periods of time are not completely clear. High-quality, publicly available field data on the long-term operational performance of PV systems are limited. In most cases, the production of field data takes many years and the technology has changed. That is why independent and high-quality laboratory data have been established in recent years to assess the long-term quality and reliability of photovoltaic modules.

PV systems are rightly regarded as low-maintenance and less susceptible to faults. These positive experiences must not be influenced by difficulties and reliability problems with additional features that may compromise the initial characteristics. For example, electronics on the module might be more susceptible to overvoltages, and this increases the risk of failures.

Comprehensive monitoring, evaluation, inspection and measurement methods detect possible defects and failures at an early stage, which can lead to performance deficits and reliability problems.

Since 2012, PV modules have been included in the EU WEEE Directive [11]. Collection and recycling of discarded end-of-life PV modules are mandatory to an extent of 80%.

2. Climatic stress factors

Solar irradiation, UV irradiation, humidity, wind, snow, rain, hail, high/low temperatures, temperature changes, salt, sand, dust and gases (O_3, NH_3, etc.). The module aging and the decrease in performance with time are complex processes that involve the interaction of several factors, namely, (a) reliability in PV, (b) service use weather conditions, (c) module materials (cell technology, encapsulant material, etc.) and design, (d) module mounting system and (e) module manufacturing process and quality control practices. Discoloration of the polymeric encapsulant is the most frequent degradation mode in hot climates. One degradation mode can be caused by a combination of various environmental stresses.

According to international report [12], at least 2% of the solar photovoltaic (PV) modules in operating solar plants will fail after 10 years of operation. As the modules are series-connected to other "healthy" modules in strings, they can lead to monetary losses as high as 27% of total income—depending on location, type of solar plant and remuneration fee.

3. HALT and HASS

Just like pulling on a chain until the weakest link breaks, HALT methods apply a wide range of relevant stresses, both individually and in combination, at increasing levels in order to expose the least capable element in the system. A highly accelerated life test (HALT) is a process that requires specific adaptation when it is applied to almost any system and assembly. A highly accelerated stress screening (HASS) is an ongoing application of combinations of stresses, defined from stress limits found empirically during HALT to detect any latent defects or reduction in the design's strength introduced during mass manufacturing.

Any stress that exposes design weaknesses that would show up in the normal environments is suitable even if theses stresses do not occur in the normal environments to which the product will be exposed. This is due to crossover effect. High reliability can be attained with careful testing that targets possible design problems, based on the physics of the failure modes, HALT testing and field data.

The highly accelerated stress test (HAST) is a process that applies increasing levels of stress (beyond the environment for which it is intended), for a short period of time, to highlight design and process weaknesses. It consists of a search for the real limits of the product based on its ability to resist environmental constraints such as random vibrations and cold and hot temperatures. This type of test is first of all focused on the very quick identification of defects. Then, corrective actions are implemented to quickly increase the product's robustness. As a result, the product is able to withstand much greater stresses than those it will experience during the normal life cycle. This method is used from the design phase to highlight design problems, also during the production phase to detect process errors. It is essential (i) to reduce product development time, (ii) to increase product reliability and (iii) to reduce after-sales service and other catastrophic youth defects in cost and brand image. It is no longer enough to measure the reliability of a product and perform a simple function test before distribution. It becomes necessary to build reliability from the upstream study around the product. Tests such as HALT and HASS allow this.

Dedicated combined environmental resources and a test time of about a week, combined with product monitoring during the tests, are the key to the success of a fast and efficient method. HALT allows you to do in a week what other environmental tests do in several months. HASS takes place during the production phase of the products. A profile is applied for 4 h (called burn-in) during which it undergoes thermal variations, random vibrations and electrical stresses. This profile allows to transform existing weaknesses on the product into measurable defects. In this way, potentially defective products are not distributed, and after-sales costs are avoided. This test takes place during the production phase of the products.

4. Useful life

The useful life corresponds to the majority of the system's lifetime. During this period, the failure rate may be:

- Increasing for mechanical elements due to wear, fatigue and corrosion.
- Constant for electronic components because there are no aging phenomena.
- Decreasing in the case of software with error correction, to improve reliability.

The age period corresponds to the failures defining the end of use of the product regardless of the type of technology. The default rate in this period is growing rapidly. During this period, products that had not been deficient during the useful period generally become deficient over a very short period.

The lifetime of PV modules is a function of a few key major field stresses such as temperature, humidity, UV light and system voltage. The accelerating factor is the relation between the time spent in the field test (or in use) and the time spent in the accelerated test. The purpose of these tests is to shorten the duration of the tests under simulated test conditions that are much stricter than actual field operating conditions, without changing the actual failure mechanisms in the field. In the AT programs, the stress tests of PV modules are performed at higher levels than the field/use stress levels along with pre- and post-characterization of materials and modules from reliability, durability and safety perspectives.

5. Levelized cost of electricity (LCoE)

The LCoE is a present value assessment of the total system costs over its lifetime and the system returns. Financial factors (the inflation or the cost of capital) are also taken into account and discounted depending on the LCoE formula in use

$$\text{LCoE}\ (\text{€}/\text{Wh}) = \frac{\text{lifecycle cost}}{\text{lifetime energy yield}} \tag{1}$$

To reduce LCoE, current research focuses on a few well-defined topics:

- Decreasing cell-to-module loss to increase module power.
- Simulation and optimization of outdoor module performance for increased energy output.
- Developing new module designs and fabrication processes for cost reduction.
- Understanding degradation mechanisms with the aim to improve module reliability and durability. PV module reliability is dependent on the quality and integrity of the process used to manufacture the module. Even small variations in material quality or manufacturing processes can impact the reliability of a component.

6. Degradation and failure

Multiple factors affect the reliability and long-term performance of PV modules. The quality and characteristics of the used materials, the interaction of the components, the manufacturing process and the local climate at the operating site play a significant role. The main degradation modes are known and their causes have been identified in most cases. To identify the main modes of degradation, accelerated aging is used with climate chambers, in situ monitoring and the variation of test parameters. The nanoindentation was used as a suitable destructive method to investigate the changes in hardness occurred during dampheat (DH) aging of components in the PV modules with a high spatial resolution [13]. Certification tests are designed to ensure a nominal level of safety and design quality, and not to indicate whether or not a product will last for its warranted lifetime; these tests are designed to ensure safety and identify infant mortality issues due to basic manufacturing quality.

The root causes of different module failures are (i) snail trails and pearl chains, (ii) cell cracks, (iii) solder joint, (iv) breakages, (v) light-induced degradation (LID), (vi) potential induced degradation (PID), (vii) hot spots and (viii) delamination. Dominant failure pathway (for modules manufactured in the early 2000s) [14]: (1) cell interconnects become more resistive, apparently due either to corrosion or fatigue of the interconnect ribbons or solder connections; (2) interconnects resistively heat, which increases the severity of temperature cycling and leads to even higher resistive heating (3) interconnect overheat or breakage leads to backsheet blackening, power loss, or other failure events. The primary underlying causes of module failures in the field were due to cell/interconnect breakage and corrosion.

In addition to failure rate, power degradation is a critical module behavior; a power degradation of a few percent has a direct effect on production over a module's lifetime, and such a small percent drop would not justify a claim under most module warranties. Subtle variations between cells, such as cell thickness, can have very large impacts to mechanical stability and reliability.

7. Predominant degradation modes

Corrosion, discoloration delamination and breakage of PV modules' encapsulant are the main modes of product degradation. Temperature and humidity are factors of PV module degradation in almost all identified degradation modes. One key factor of reducing the costs of photovoltaic systems is to increase the reliability and the service lifetime of the PV modules. Models can help to overcome the long-term experiments' obstacle in order to study PV module degradation under real conditions. To increase the reliability and the service lifetime of the PV modules, it is necessary to reduce the costs of PV systems.

In addition to degradation analysis, the stress tests available today are very effective at screening for PV module defects that cause severe degradation or safety issues such as bad solder joints or a poorly adhered junction box.

8. Environmental factors influence the degradation

Even bird droppings can significantly reduce the output of a photovoltaic module. However, many causes of faults are not visible to the naked eye. In such cases, special devices such as a thermal imaging camera are required for localization.

With a thermal imaging camera, the following defaults in the modules can be detected: (a) production failures, (b) damage such as cracks, (c) faulty power connections and connections, (d) contamination and shading, (e) defective cables and (f) inverter damage.

9. PID and LID

PID and LID are two different kinds of induced degradation of PV modules. In the first case, *potential induced degradation* (PID) is conducted by high voltages and the other *light-induced degradation* (LID) conducted by sunlight (real or simulated).

PID is the "new disease" of the PV module; highlighted in 2010, it begins to affect more and more photovoltaic modules. The first symptom of this phenomenon is a rapid and unexplained degradation of power. This decrease in module efficiency, which can reach more than 20% in a few months, is neither due to conventional module aging nor to improper module installation. Individual modules in PV systems are often connected in series to increase the system voltage. The potential difference of the chain thus formed can sometimes reach a few hundred volts [15]. In order to protect people from electric shock, all metal structures of the modules are often grounded. Thus, leakage currents can occur due to a lack of insulation between the structure and the active layers (PV cells) [16]. This phenomenon can lead to polarization that can degrade the electrical characteristics of photovoltaic cells. This phenomenon known as potential induced degradation (PID) is characterized by the progressive degradation of the performance of crystalline silicon-based photovoltaic modules due to the presence of an induced electric current at the very heart of the module [17, 18]. Hacke showed that PID was more common in humid climates than in hot and dry environments [19]. Schütze et al. go in the same direction by showing that leakage currents increase with humidity [15]. In his study, a ramp voltage of −600 V at sunrise and 0 V at sunset is applied between the metal structure and the contact of a module composed of 60 cells. His experience has shown that leakage current increases with humidity. The various studies have shown that the main factors favoring PID are the voltage of the system in which the module is used, humidity and operating temperature. PID is triggered by a combination of high temperature and relative humidity. Furthermore, some degradation mechanisms are interdependent. The PID process that occurs more often in p-type cells is a PID of the shunting type, sometimes denoted by PID-s. Other types of PID in p-type cells are the dissolution of the antireflective coating and the corrosion of the cell fingers [20]. Quality control during the module manufacturing process is essential to ensure a good performance in the long run.

ID usually occurs when modules are in strings operating at high voltages (near 1000 V_P, but not only), combined with very warm and humid weather. Dust and glass degradation (releasing sodium ions) may catalyze the PID phenomenon. Potential induced degradation (PID) is an effect that affects some PV modules with crystalline Si cells; the degradation of performance that can occur after a few years can arrive to 30% or more.

Light-induced degradation (LID) has been identified to be a critical issue for the long-term stability of solar cells and modules from boron-doped silicon substrates. Besides the well-known LID of excess charge carrier lifetime within Czochralski-grown silicon substrates induced by the activation of the boron-oxygen complex, significant performance degradation has been observed also for certain multicrystalline silicon (mc-Si) solar cells and modules [21]. This degradation is significantly more pronounced at elevated temperatures and, therefore, referred to as LeTID for "light and elevated temperature-induced degradation." If not controlled, LeTID can induce a decrease of conversion efficiency by more than 10%, particularly for solar cells with dielectrically passivated surfaces. LeTID needs therefore to be suppressed by adapted cell processing.

For the younger installations (less than 10 years of operation), the most observed degradation modes are hot spots and internal circuitry discoloration (both related to electric interconnections), encapsulant discoloration, broken cells and PID (**Figure 2**).

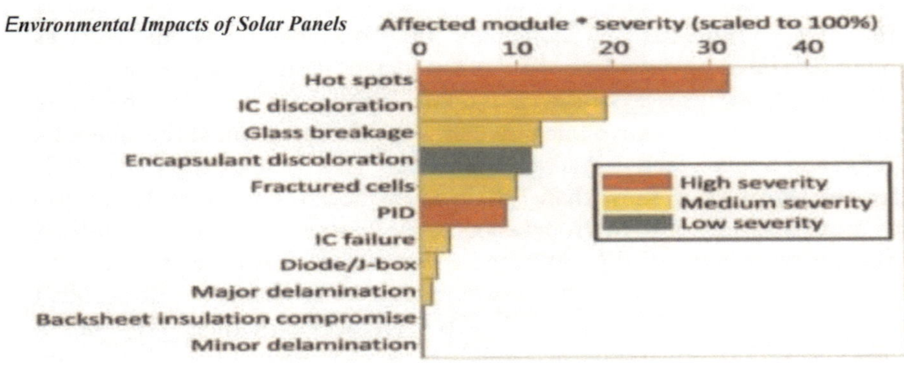

Figure 2.
Pareto chart of the most significant degradation modes for systems installed in the last 10 years. PID is ranked as a high-severity degradation mode. Interconnection failures, which include hot spots and internal circuitry (IC) discoloration, also play a large role in degradation. Image adapted from [22, 23], with kind permission.

The LID for crystalline silicon modules varies between 0.5 and 3%; some modules have a loss of up to 5%. Manufacturers using n-type silicon cells have no loss of LID. Manufacturers take this into account by considering a power loss of 3% during the first year of the module warranty [24].

10. Failures in the electric interconnections

Interconnection failures are generally caused by thermo-mechanical fatigue. Because of the different coefficients of thermal expansion of the materials in the module sandwich (e.g. cells, encapsulant, glass, back sheet), thermal variations result in a displacement of solar cells in the module. The same effect can occur when the module is exposed to mechanical stresses, e.g. due to wind or snow loads.

The reliability of ribbons has gained increasing attention in the last few years. In [23] how the module performance is affected when one or more ribbons are broken was investigated. It was quantified, by experimental measurements, the actual impact on the electric performance when one or more ribbons connecting adjacent cells are disconnected. In [23] a simplified electrical model in LT-SPICE that is able to reproduce experimentally measured IV characteristics and the variation in the module's electrical parameters induced by the breakage of the cell interconnect ribbons were proposed.

11. Defects

A defect is everything in a PV module that is not as it is expected to be. A large number of defects are known in PV c-Si modules. Some of these can be identified by visual inspection [25, 26]:

- Broken glass by volley or mechanical accidents
- Delamination (busbars, EVA foil)
- Broken or deformed frames
- Brownish color caused by hot spots
- Charred electrical connection box
- Defects in connectors
- Environmentally altered surfaces
- etc.

12. Rate of defective panels

The rate of defective panels leaving the assembly lines—evaluated thanks to returns and complaints from customers—is well below 0.1%. Highly demanding quality standards have been defined by the IEC, an international body that sets design and quality standards.

The youth period refers to early failures due to design problems (incorrect sizing of a component, etc.) or production (deriving from a process of manufacturing). The default rate is decreasing in this period. In the case of modules' photovoltaic, youthful failures can be eliminated before delivery to the customer by practicing debugging. Burn-in consists of turning on the components to be delivered under conditions that may reveal failure modes. This practice is expensive and the failure rate at delivery is equal to that at the beginning of the useful period. Many manufacturers do not perform this burn-in on their products for reasons of cost. In this case, a warranty period is put in place during which the manufacturer undertakes to change or repair the defective product. For example, for photovoltaic modules, manufacturers guarantee them for an average of 5 years for mechanical failures (unrelated to the power delivered by the modules). In reliability studies, defects that appear during this warranty period are not taken into account and are mainly concerned with the useful life of the product.

13. Failures

Typically failures of products are divided into the following three categories: *infant failures, midlife failures* and *wear-out failures*. During initial exposure to light, crystalline silicon modules generally undergo a permanent reduction in output power emissions. This phenomenon is called light-induced degradation or LID. On average, the LID for crystalline silicon ranges from 0.5 to 3% due to traces of oxygen included in the molten silicon during Czochralski process. Manufacturers take into account a 3% power loss during the first year of the module's warranty. Be aware of LID degradation [27]. Crystalline p-type boron-doped silicon solar cells generally exhibit a degradation of conversion efficiency during the first hours of exposure to the sunlight. Understanding light-induced degradation of c-Si solar cells is associated with the formation of the well-known boron-oxygen complex which acts as a harmful defect and reduces the minority carrier diffusion length accordingly. LID is therefore related to both, boron and oxygen concentrations.

For thin-film PV modules, there are far fewer experiences accumulated in the past years than for crystalline Si PV modules.

Failure modes are directly related to module material sand; hence, they continuously evolve with the material adjustments made by module manufacturers.

Infant-mortality failures occur in the beginning of the working life of a PV module. Flawed PV modules fail quickly and dramatically impact the costs of the module manufacturer and the installer because they are responsible for these failures.

Failures occurring in the midlife of PV modules are described in a study of DeGraff [28]. **Figure 3** shows the analysis and statistics of specific field failure distribution. The predominant PV module failures are delamination, cell part isolation due to cell cracks, and discoloring of the laminate.

Module failure. The so-called ribbon kink (between the cells and the joint between the cell interconnect ribbon and the string interconnect) is prone for fatigue breakage. The possible causes are poor soldering → disconnections. A too intense deformation during the fabrication of the ribbon kink between the cells mechanically weakens the cell interconnect ribbon. A narrow distance between the cells promotes cell interconnect ribbon breakage. Physical stress during PV module transportation, thermal cycle and/or hot spots by partial cell shading during long-term PV system operation forces mechanical weak ribbon kinks to break.

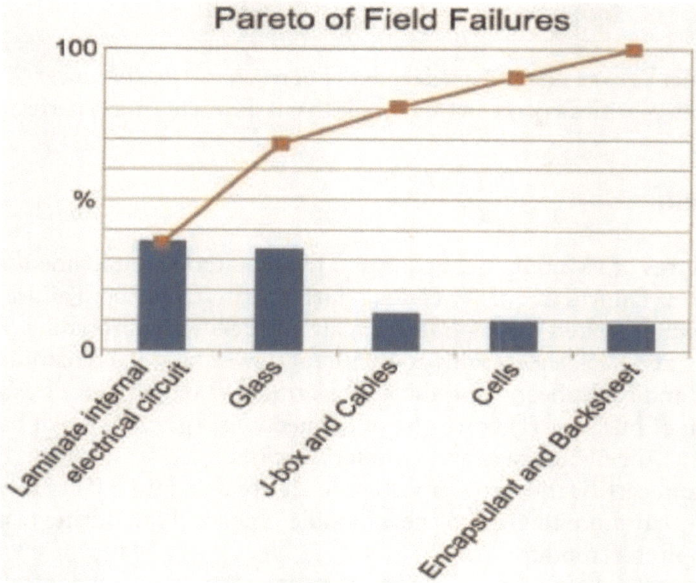

Figure 3.
Analysis and statistics of specific field failures (with kind permission of SunPower Corp.)

13.1 External causes

Some failures are typically difficult to define as a PV module failure or as a failure of the contractor, of the installer or the system designer or even for other reasons. The specific field failures are their analysis and statistics (**Figure 3**).

13.2 Clamping

A relatively often seen failure in the field is glass breakage of frameless PV modules caused by the clamps. Major problems caused by glass breakage are electrical safety issues, because the insulation of the modules is no longer guaranteed, in particular in wet conditions, and because glass breakage causes hot spots, which lead to overheating of the module.

13.3 Transport and installation

Even in well-designed transport containers, the cells of PV modules may crack during "normal" transport. The cause of cell breakage is much more difficult to decide. Only an electroluminescence image or a lock-in thermography image can reveal the damage.

13.4 Quick connector failure

Quick connector failure—in the most cases—is not considered a PV module failure.

13.5 Lightning

A defective bypass diode caused by a lightning strike is caused by an external source, for which the module is not designed.

13.6 Defect

Much wider than a failure, the defect does not result in a loss of power or safety of a PV module but specifies a part of a PV module that is different from a perfect PV module.

14. Field failures

Field failures of PV equipment can stem from materials, fundamental product design flaws or failures in quality control during manufacturing. Failure modes in the field which occurred during operation are (a) yellowing/browning of encapsulants and back sheets with and without power loss, (b) delamination of encapsulant and back sheet, (c) bubble formation, (d) oxidation of busbars, (e) discoloration of busbars, (f) corrosion of connections, (g) cracking of back sheet, (h) hot spots, (i) cell breakage and (j) microcracks.

Testing and certification are important to assure a certain quality level, taking into account that more than 1/3 of new module types still fail during testing for certification in the laboratories.

Check the module for damage due to transportation before the installation. Do not use or install damaged modules. Damaged modules may cause fire or electric shock, resulting in property damage, fire and or death.

LEDs are gaining an increased market share in applications requiring light in the UV spectrum [29]. This trend is driven by big technological advancements and various advantages of LEDs compared to the conventional UV light sources. These advantages include a higher efficiency, a longer lifetime, more constant radiance and less heat generation [30].

It was imperative to evaluate the option of using UV LEDs in PV module and component degradation. The UV LEDs have proven their long lifetime and slow degradation at room T and RH. However, also polymeric materials only degrade slowly in those conditions. This emphasizes the need to apply UV light and other stress factors at the same time in accelerated aging and degradation testing [31].

15. The most common failures of PV modules

Four major materials are assembled to realize a PV module: glass, metals, polymers and solar cells/active semiconductor thin-film layer.

Aging and failure mechanisms seen over the past several decades have been documented over a wide range of power plant locations and material sets (**Figure 4**).

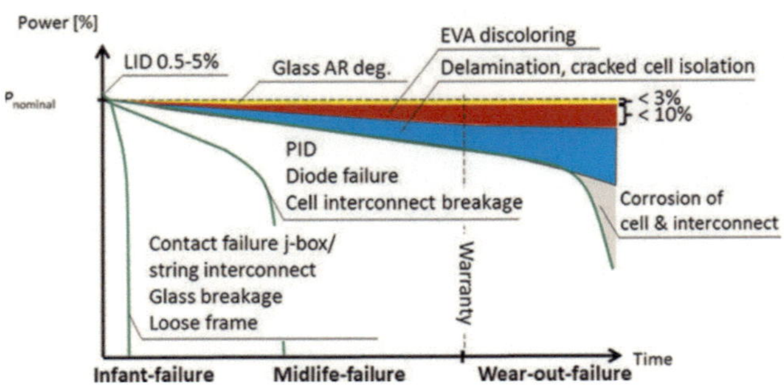

Figure 4.
Three typical failure scenarios for wafer-based crystalline photovoltaic modules are shown. Definition of the used abbreviations: LID, light-induced degradation; PID, potential induced degradation; EVA, ethylene-vinyl acetate; jbox, junction box [32]. With kind permission of IEA.

indicates leading PV module aging and failure mechanisms that occur as infant mortalities, midlife failures and wear out. Laminates containing EVA foils with a systematic variation of the additive formulation, i.e. the crosslinking agents, ultraviolet (UV) absorber, hindered amine light stabilizers and antioxidants, were subjected to UV aging. Failures are the following: EVA discoloration; cell cracks; snail marks; loss of adhesion of the back sheet; interconnection tapes of the cell and disconnected wires; junction box failure; frame failure; burn marks; potential induced degradation; defective bypass diodes; failure of thin-film modules, such as microarcs at hot spots of bonded connectors; shunts and front glass failure; and rear contact degradation [32].

16. Ethylene-vinyl acetate (EVA)

One of the most overt degradation mechanisms for PV modules is the discoloration of the ethylene-vinyl acetate (EVA) or other encapsulation materials (considered as an esthetic issue). EVA is usually formulated with additives, including UV and thermal stabilizers. But if the choice of additives and/or their concentrations are inadequate, the EVA may discolor.

Delamination may be more likely at the interface between EVA and the solar cell, because the interfacial strength may initially be more limited there than at the EVA/glass interface. On the other hand, UV degradation and subsequent embrittlement may limit the long-term adhesion of interfaces exposed to the sun.

It is quite common to see symmetric patterns and sometimes multiple rings based on the effects of limited chemical diffusion, both into and out of EVA and the existence of multiple chemical pathways that produce similar chromophore species.

It was possible to show a link between changes in mechanical properties with both the transient temperature and the degree of long-time thermal aging [33].

The overall EVA degradation is higher in the Indonesian equatorial zone, in the African semi-arid zone and in the Chinese cold desert (**Figure 5**).

Results from environmental degradation simulations highlighted the effect of climatic zones on degradation, with higher degradation in hotter, warmer and more irradiated zones. The overall EVA degradation is higher in the Indonesian equatorial zone, in the African semi-arid zone and in the Chinese cold desert [34]. With kind permission of the authors.

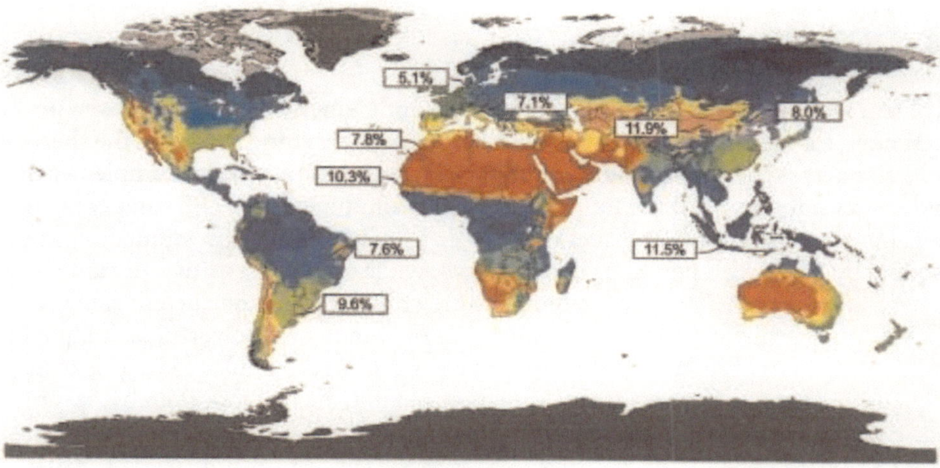

Figure 5.
Köppen's map and overall EVA degradation, calculated after a 20-year exposure in different climatic zones.

Glass can be a source of module reliability problems if not properly fabricated. Glass reliability issues can include brittle failure due to mechanical and thermal stresses, surface weathering, lamination adhesion, transparent conductive oxide (TCO) adhesion, moisture ingress and antireflective coating durability.

17. Failure analysis

The search for physical-failure root causes is aimed at breaking through any technology barrier in each stage of chip development, package development, manufacturing process and field application, offering the real key to eradicate the error. Testing provides us with information on the electrical performance; FA can discover the detractors for the poor performance [15].

Failure modes in the field which occurred during operation are analyzed [31]: microcracks, cracking of back sheet, corrosion of connections, hot spots, cell breakage, delamination of encapsulant and back sheet, oxidation of busbars, yellowing/browning of encapsulants and back sheets with and without power loss, bubble formation and discoloration of busbars.

Most of the failure modes observed are polymer defects such as delamination, yellowing (there are no obvious results that show a direct relationship to power losses in the event of yellowing) and browning, bubbling and cracking of polymers used for encapsulants and back sheets. Oxidation and discoloration of busbars and corrosion of connectors are also easily detected without special measuring equipment; they lead to increased resistance in series. Microcracks, cell failure and hot spots have a greater impact on performance [31]. The detection of these defects requires IR cameras.

Modules under test both in the field and during qualification testing have exhibited several types of failure resulting from moisture intrusion. Poorly designed cell assemblies can damage or break the cells.

Failures observed during the analysis are the following: (i) moisture intrusion, (ii) differential thermal expansion, (iii) plus UV exposure, (iv) wind loading, (v) off-track charring and (vi) hot spot. The types of failures resulting from moisture intrusion are (1) cell strings shorting to ground; (2) terminals shorting to ground; (3) high ground fault currents, preventing inverter from coming on; (4) modules filling with water; (5) corrosion; and (6) degradation of optics.

18. New techniques for failure analysis

Dark lock-in thermography (DLIT) has recently emerged as a new powerful technique for failure analysis of PV modules; it provides information on the thermal behavior across the whole module at high spatial resolution, besides complementing and overcoming some of the limitations of electroluminescence (EL) and conventional passive infrared (IR) cameras. Its basic principle consists in stimulating the module by applying a periodic current signal, while a highly sensitive IR camera detects and measures the surface temperature stimulated by the current signal.

Scanning acoustic microscopy[2] (SAM) has proven to be a powerful technique for the investigation of failures and defects in solar cells and modules [35]. This technique is needed to advance the assessment of module quality and, thereby, improve

---

[2] SAM is a non-invasive and non-destructive technique that can be used to manage the internal features of a specimen, in particular its interfaces. SAM is a superior tool to detect delamination even of submicron thicknesses (down to 100 nm).

their long-term performance. The PV modules were placed flat into a water tank and scanned.

PV modules (20 × 20 cm^2) were prepared by laminating a layer of polymer-based backsheet foil (BS) and two layers of ethylene-vinyl acetate (EVA) sandwiching a monocrystalline silicon cell and a transparent glass cover.

The very fine resolution achieved with SAM allowed a clear visualization of the location and shape of the air pockets detected on the rear side of the module. These air pockets seem to form only on the silver pads next to a ribbon, which may indicate that the EVA did not adhere completely to the silver pads, thus giving rise to air gaps. In acoustic micrographs, air pockets and delaminated areas are easily detected as very bright features because air-filled interfaces reflect almost all incoming sound.

The depth profile analysis indicated that SAM can be applied to detect thickness variations within the BS and encapsulant layers. SAM has proven to be a very sensitive technique to detect interfacial faults, such as cracks on the solar cell, bubbles and air pockets at different interfaces. A correlation between SAM and DLIT images with regard to the encapsulation homogeneity and adhesion defects could be verified, suggesting both techniques are complementary to study defects of this nature. Therefore, SAM is a reliable and complementary technique to DLIT and EL in investigating PV modules [10].

19. Influence of temperature

Temperature is extremely significant to the PV module degradation process, especially hot spots, encapsulant bleaching, delamination failure on interconnections, corrosion, discoloration and bubbles on the panel's surface.

Crystalline silicon PV cells/module degradation—exposed under temperature and heat effect—has been investigated. This revealed:

- Delamination of encapsulant and back sheet.
- Bubble formation, oxidation of busbars, yellowing/browning of encapsulants and back sheets with and without power loss.
- Discoloration of busbars.
- Corrosion of connections.
- Cracking of back sheet.
- Hot spots, cell breakage and microcracks are the dominant modes of degradation.

Temperature is responsible for most of the chemical reactions and extremely significant to the PV module degradation process, especially hot spots, encapsulant bleaching and delamination failure on interconnections, corrosion, discoloration and bubbles on the panel's surface.

The leakage current increases rapidly with increasing number of hot spots. The effect of discoloration causes loss of transmittance of the encapsulant EVA and reduces the photocurrent density (J_{ph}) owing to a decrease of absorption and therefore the power loss. Discoloration does not affect the fill factor (FF) and (Voc) more, but the corrosion causes a decrease of the PV module maximal power (P_{max})

Accelerated Stress Test	Failure mode	International regulation
Thermal cycling	Broken interconnections Cell cracks Solder bond failures Module open-circuit Potential for arcing Junction box adhesion	IEC 61215
Dampheat	Corrosion Delamination Encapsulant loss of adhesion & elasticity Junction box adhesion Inadequate edge deletion	IEC 61215
Humidity freeze	Delamination Junction box adhesion Inadequate edge deletion	IEC 61215
UV exposure	Delamination Encapsulant loss of adhesion & elasticity Encapsulant & backsheet discoloration Ground fault due to backsheet degradation	IEC 61215
Static Mechanical Load	Structural failures Broken glass Broken interconnect ribbons Broken cells Solder bond failures	IEC 61215
Dynamic Mechanical Load	Broken glass Broken interconnect ribbons Broken cells Solder bond failures	IEC 61215
PID test	Potential-induced degradation	IEC62804 [59]

Table 1.
Summary of the main accelerated tests for c-Si PV modules (Adapted from [37]).

	Qualification	Comparative	Lifetime
Purpose	Minimum design requirement	Comparison of products	Lifetime estimation
Quantification	Pass/fail	Relative	Absolute
Climate	Not differentiated	Differentiated	Differentiated

Table 2.
Three types of accelerated tests employed in the PV industry.

when the delamination of the PV module reduces the thermal conductivity locally and hence increases the temperature of the cell [36] (**Tables 1** and **2**).

20. Accelerated tests

The *International Photovoltaic Quality Assurance Task Force* (PVQAT) gathers the efforts of many research groups worldwide in order to improve the understanding of the reliability of PV modules. One goal is to propose a climate-specific test protocol.

21. Qualification tests

The qualification tests have been essential to identify materials that lead to early failures in the field and to achieve a high reliability for PV modules. However, they present some limitations if one wants to interpret them as reliability-predictive tests. Presently, PV modules are only going through qualification tests (such as performed according to the IEC 61215 for crystalline silicon-based modules) which provide limited information as (i) they do not provide information on the maximum lifetime that can be expected and (ii) they do not consider climate specificities [39].

Testing and certification of a PV module focuses on verifying that the fundamental design requirements have been fulfilled. Reliability testing increases confidence in production quality and usually takes less time and costs less than durability tests [40].

The purpose of the qualification tests is to screen new designs and new production runs for susceptibility to known failure mechanisms: (1) The tests do not establish field lifetimes; (2) cell assemblies and complete modules are tested separately.

22. Photometric testers

Measurements with photometric testers give additional hints on defect PV modules (defect bypass diodes, broken or delaminated connectors and wires). These inspections of modules often can be done while these modules are still installed. However, analysis with photometric testers addresses the modules as a whole and cannot localize the defect within the module; defects may be localized by thermographic cameras.

All critical areas that are detectable via the IR camera at long wavelength could as well be detected with the near-infrared (NIR) camera. However, NIR images are more suitable for the defect analysis since they have higher local resolution [41].

Typically the NIR images show an affine distortion. This has to be corrected carefully to an orthoimage. Furthermore NIR image is brighter at the center than in the outer regions. This must be numerically compensated.

23. Combining up to three stress factors targeting specific mechanisms

The necessity for combining multiple stress factors into fewer tests has been recognized for some time, as demonstrated by efforts made by various other groups [42–45]. While important results have been produced, these efforts are typically limited to combining up to three stress factors targeting specific mechanisms. In this work, a combined-accelerated stress testing (C-AST) capability has been developed which combines multiple stress factors from the natural environment including light, humidity, temperature, rain, mechanical loads and voltage stress. The test has been developed with the philosophy that it is agnostic to a priori known failure mechanisms such that mechanisms in new module designs or materials may be identified before deployment. C-AST is applied here to mini-module specimens to allow for possible adverse material interactions.

The test includes six 4-cell p-PERC mini-modules with three back sheets based on polyvinylidene fluoride, polyamide and polyvinyl fluoride, as well as two types of polyethylene-co-vinyl-acetate (EVA), one UV blocking and one UV transparent. The experiment demonstrated C-ASTs' ability to identify backsheet cracking and other failure mechanisms such as UV-induced degradation, solder bond breakage, minor corrosion and cell cracking.

24. Encapsulants

Encapsulation materials are of critical importance for long-term reliability and safety of PV modules. Only months in the field may lead to drastic power output degradation, for example, due to potential induced degradation (PID), and also in the longer run adhesion and discoloration issues (hot spot) can reduce power output or even may lead to critical electrical safety issues. These findings are not sufficiently covered by IEC 61215 and other standard testing. Additional and different types of tests are required for an in-depth understanding of the encapsulation material's impact on PV module performance [46]. PID can be critical for many of the solar parks built in the early phase of PV deployment (2000–2012); however, it can still occur in modern modules when lower encapsulants are used in PV manufacturing, a choice often driven by the need for cheap solar electricity.

25. Key performance indicators

Typical key performance indicators (KPIs) used in solar industry for assessment of the performance of solar modules and balance-of-system (BOS) components were investigated in [47]. From the point of view of an investor in PV systems, it is important to have a comprehensive insight into the reliability of solar modules. Since reliability and longevity of such components are crucial for the long-term financial success of the project, corresponding KPIs are desired. To this end solar industry has traditionally provided KPIs for both solar modules and inverters on the base of relating the number of failed (claimed) product to the number of installed (or sold) product. The denominator of these KPIs may be referred to as either related to the same period as the numerator or to a cumulated number of installations. The study [48, 49] has shown that currently used KPIs are not sufficient when comparing the ability of solar manufacturers to produce reliable solar modules or inverters.

Presently, PV module manufacturers typically guarantee 80% of the nominal power for 25 years. If we assume a linear power degradation rate, this corresponds to an annual degradation rate D of 0.8%/year.

To recover the losses, a continuous repowering would be necessary, substituting the modules that fail with new ones. Failures in modules lead to hot spots that can be detected by visible (VIS), IR thermography and electroluminescence (EL) images. This inspection, mainly manual, is rapidly reducing costs by the increasing penetration of drones, which will very probably become soon the state of the art [50–53].

26. Metal-wrap-through (MWT) concept

The metal-wrap-through (MWT) concept, which was developed in 1998 by Kerschaver et al. [54], represents an alternative approach for reducing shading while at the same time ensuring reliable measurability and purely rear-side contacting. MWT solar cells differ from conventional solar cells in that the front-side contact finger is not routed to busbars, but to the rear side via metallized vias. This makes it possible to connect the cells on the rear side only, but the cell structure otherwise corresponds to that of conventional solar cells.

Critical points such as the recombination in the area of the vias, the influence of the rear n-contacts to the series resistor as well as the arrangement of the rear contacts for a reliable and low-loss external circuit are to be dealt with in simulations and experiments. The selection of suitable passivation layer systems as well as the reliable insulation of the vias and the rear side n-contacts against the p-base is also the central aspects regarding the minimization of leakage currents. Another important factor for reliable module integration is the behavior of the cells when negative voltage is applied externally.

This operating state can occur during partial shading of photovoltaic modules and can lead to local overheating within the cell if the current flux is not controlled. Investigations into the backward behavior of the developed solar cells are, therefore, of great importance for the long-term stability of the solar cells and the modules produced from them.

The reliable connection of the front contact grid with the rear contact surface through vias is a central task in the development of MWT solar cells.

The efficiency of solar cells available in practice is reduced by a multitude of further loss mechanisms [55] and depends both on the quality of the materials used and on the available technologies. Predictions of the practical reachable maximum are, therefore, highly error-prone.

High reliability: service life between 10 and 20 years, depending on location installation.

27. Perspectives

The continuous evolution in cell technologies and the implementation of new encapsulation materials will lead to the appearance of new degradation modes, which might require years before being detected in the field. The case of the so-called light and elevated temperature-induced degradation (LeTID) effect that was observed in PERC cells only in 2015 is an example. Also new applications of PV such as building-integrated photovoltaics (BIPV) lead modules to operate in new conditions (resulting, e.g. in higher module temperature excursions) and employ new materials, which may result in a different evolution of the typical degradation modes as well as in the appearance of new degradation modes.

With regard to new cell technologies, we expect that in passivated emitter and rear cells (PERC), where the only differences with respect to standard p-type cells lay in the rear surface, the mechanism of PID is in general the same as for the standard p-type cells. It is expected therefore that the study of PID mitigation strategies [23] is applicable to PERC cells as well. The mechanism of PID in silicon heterojunction (SHJ) cells is instead very different in that it does not consist of cell hunting, but corrosion of the transparent conductive oxide (TCO) layer (similar to thin-film modules) requires a different study.

28. Conclusions

The reliability of PV power plants and modules has been—and will continue to be—an issue for investors, owners and utilities.

A global network is required to improve the quality and reliability of PV systems and components by collecting, analyzing and disseminating information on their technical and financial performance.

PV will utilize new materials, manufacturing methods, module and systems designs in order to lower costs and hopefully increase or maintain reliability.

New module materials and constructions and new system concepts are necessary for increased reliability and lifetime [56].

Author details

Titu-Marius I. Băjenescu[1,2]

1 Military Technical Academy of Romania, Bucharest, Romania

2 Technical University of Republic of Moldova, Chişinău, Moldova

*Address all correspondence to: tmbajenesco@gmail.com

References

[1] Chapin DM et al. A new silicon p-n junction photocell for converting solar radiation into electrical power. Journal of Applied Physics. 1954;**25**:676

[2] Becquerel AE. Memoire sur les effects d'electriques produits sous l'influence des rayons solaires. Comptes Rendus de l'Académie des Sciences. 1839;**9**:561-567

[3] Goetzberger A et al. Solar cells: Past, present and future. Solar Energy Materials & Solar Cells. 2002;**74**:1-11. Available from: http://193.140.122.139/solar/resources/e-books_and_papers/web-editions%20-%20goetzberger%20-%20Solar%20cells,%20past,%20present,%20future.pdf

[4] Wolf M. High efficiency silicon solar cells. In: Proceedings of the 14th IEEE Photovoltaic Specialists Conference; San Diego; 1980. p. 674

[5] Chittick RC et al. The preparation and properties of amorphous silicon. Journal of the Electrochemical Society. 1969;**116**:77

[6] Spear WE, LeComber PG. Investigation of the localised state distribution in amorphous Si films. Journal of Non-Crystalline Solids. 1972; **8**:727

[7] Carlson D, Wronski C. Amorphous silicon solar cell. Applied Physics Letters. 1976;**28**:671

[8] Essakiappan S et al. Current status and future trends in solar technology—A comparative study of Texas and California. Technical Report: TR-2010-ECE689-Fall Group No. 1; December 10, 2010

[9] Kurtz S. Reliability and Durability of PV Modules. 2017. DOI: 10.1002/9781118927496.ch44

[10] European Commission. Directive 2012/19/EU of the European Parliament and of the Council on Waste Electrical and Electronic Equipment (WEEE); July 4, 2012

[11] Köntges M et al. Review of failures of photovoltaic modules. Report IEA-PVPS, T13:01; 2014

[12] Gray KA, Paschkewitz JJ. Next Generation HALT and HASS, Robust Design of Electronics and Systems. John Wiley & Sons, Ltd; 2016

[13] Mansour DE, Swientek F, Kaaya I, Philipp D, Bauermann LP. Nanoindentation analysis of PV module polymeric components after accelerated aging. In: Proceedings of the 35th EU PVSEC 2018; September 24–28; Brussels; 2018. pp. 1333-1336

[14] Owen-Bellini M, Hacke P, Spataru S, Miller DC, Kempe M. Combined-accelerated stress testing for advanced reliability assessment of photovoltaic modules. In: Proceedings of the 35th EU PVSEC 2018; September 24–28; Brussels; 2018. pp. 1101-1105

[15] Schütze M, Junghänel M, Friedrichs O, Wichtendahl R, Scherff M, Müller J, et al. Investigations of potential induced degradation of silicon photovoltaic modules. In: 26th European Photovoltaic Solar Energy Conference; Hamburg, Germany; September 5–9, 2011

[16] Schütze M et al. Laboratory study of potential induced degradation of silicon photovoltaic modules. In: 37th IEEE PVSC; 2011

[17] Pingel S et al. Potential induced degradation of solar cells and panels. In: 35th IEEE PVSC; 2010

[18] Berghold J et al. Potential induced degradation of solar cells and panels. In: 25th EUPVSEC; 2010. pp. 3753-3759

[19] Hacke et al. System voltage potential-induced degradation mechanisms in PV modules and methods for test. In: 37th IEEE PVSC; 2011

[20] IRENA and IEA-PVPS. 2016

[21] DeGraaff D, Lacerda R, Campeau Z. Degradation mechanisms in Si module technologies observed in the field; their analysis and statistics. Presentation at PV Module Reliability Workshop 2011. NREL, Denver, Golden, USA; 2011. Available from: http://www1.eere.energy.gov/solar/pdfs/pvmrw2011_01_plen_degraaff.pdf

[22] Yole Developpement. UV LEDs—Technology, Manufacturing and Application Trends Report. 2018

[23] Muramoto Y, Kimura M, Nouda S. Development and future of ultraviolet light-emitting diodes: UV-LED will replace the UV lamp. Semiconductor Science and Technology. 2018;**29**(8): 0848004

[24] Mitterhofer S, Jankovec M, Topič M. Using UV LEDs for PV module aging and degradation study. In: Proceedings of the 35th EU PVSEC 2018; September 24–28; Brussels; 2018. pp. 1323-1327

[25] Ferara C, Philipp D. Why do PV modules fail? Energy Procedia. 2012;**15**: 379-387

[26] IEA. Review of failures of photovoltaic modules. Report IEA-PVPS T13-01:2014; 2014

[27] Badiee A, Ashcroft IA, Wildman RD. The thermo-mechanical degradation of ethylene vinyl acetate used as a solar panel adhesive and encapsulant. International Journal of Adhesion and Adhesives. 2016;**68**: 212-218

[28] Gagliardi M, Paggi M. Long-term EVA degradation simulation: Climatic zones comparison and possible revision of accelerated tests. Solar Energy. 2018; **159**:882-897

[29] Conference 35th EUPVSEC Key Performance Indicators and PV Module Reliability Problem; September 2018

[30] Mesquita LV, Mansour DE, Philipp D, Bauermann LP. Scanning acoustic microscopy as a non-destructive method for the investigation of PV module components. In: Proceedings of the 35th EU PVSEC 2018; September 24–28; Brussels; 2018. pp. 1318-1322

[31] Hounkpatin GF, Kounouhéwa BB, Agbomahéna M, Madogni VI. Degradation of crystalline silicon photovoltaic cells/modules under heat and temperature effect. Physical Science International Journal. 2018;**19**(1):1-12

[32] Virtuani A, Annigoni E, Martins A, Niquille X, Gnocchi L. PV Modules: Reliability and Performance Prediction. Available from: https://pvlab.epfl.ch/page-131402-en.html

[33] Breitenstein O et al. Lock-in Thermography-Basics and Use for Evaluating Electronic Devices and Materials. Springer; 2010

[34] Koehl M, Hoffmann S, Weiss K-A. Combined stress factor testing of PV modules. In: Proceedings of 29th European Photovoltaic Solar Energy Conference and Exhibition; 2014

[35] Masuda et al. Sequential and Combined Acceleration Tests for Crystalline Si Photovoltaic Modules. 2016

[36] Gambogi W. Sequential stress testing to predict photovoltaic module durability. In: 7th World Conference on Photovoltaic Energy Conversion; 2018

[37] Tamizhmani G. Long-term sequential testing (LST) of PV modules.

In: Photovoltaic Module Reliability Workshop; 2012

[38] Pingel S, Janke S, Stannowski B, Fechner S, Podlowski L. Advanced testing of PV module encapsulants. In: Proceedings of the 35th EU PVSEC 2018; September 24–28; Brussels; 2018. pp. 1346-1351

[39] Kleiss G. Key performance indicators and PV module reliability. In: Proceedings of the 35th EU PVSEC 2018; September 24–28; Brussels; 2018. pp. 1355-1359

[40] dos Reis Benatto GA et al. Luminescence imaging strategies for drone-based PV array inspection. In: Proceedings of the 33rd EUPVSEC; Amsterdam; 2017. pp. 2016-2020

[41] Koch S et al. Outdoor electroluminescence imaging of crystalline photovoltaic modules: Comparative study between manual ground-level inspections and drone-based aerial surveys. In: Proceedings of the 32nd EUPVSEC; Munich; 2016. pp. 1736-1740

[42] Lanz M et al. Drone-based assessment of cleaning effects on PV installations. In: Proceedings of the 32nd EUPVSEC; Munich; 2016. pp. 1960-1963

[43] Kerschaver EV, Einhaus R, Szlufcik J, Nijs J, Mertens R. A novel silicon solar cell structure with both external polarity contacts on the back surface. In: Proceedings of the 2nd World Conference on Photovoltaic Energy Conversion (Vienna); 1998. pp. 1479-1482

[44] Swanson R. Approaching the 29% limit efficiency of silicon solar cells. In: Proceedings of the 31th IEEE Photovoltaic Specialists Conference (PVSC); Orlando, FL, USA. Piscataway: IEEE; 2005. pp. 889-894. DOI: 10.1109/PVSC.2005.1488274

[45] Jordan DC, Kurtz SR. Photovoltaic degradation rates—An analytical review. Progress in Photovoltaics: Research and Applications. 2011;**21**: 12-29. DOI: 10.1002/pip.1182

[46] T.-M. I. Băjenescu, Advances and trends in photovoltaics, Electrotehnică, Electronică, Automatica vol. 61(2013), nr. 1, p. 7-13

[47] T.-M. I. Băjenescu, Present and future of photovoltaics, Electrotehnică, Electronică, Automatica vol. 63(2015), nr. 1, p. 31-38

[48] Băjenesco T-MI. La photonique verte. Electrotehnică, Electronică, Automatică. 2017;**65**(1):7-12

[49] Băjenesco TI. Problèmes de la Fiabilité des Composants Électroniques Actifs Actuels. Paris & Arm, Suisse: Masson; 1980. ISBN: 22225699607

[50] Băjenescu T-MI. Achievements and trends in photovoltaics. Electrotehnică, Electronică, Automatică. 2017;**65**(4): 123-127

[51] Băjenescu T-MI. State of the art of photovoltaics. Meridian Ingineresc. 2016;**4**:13-23

[52] Băjenescu T-MI, Bâzu M. Reliability of Electronic Components—A Practical Electronic Systems Manufacturing. Berlin and New York: Springer; 1999. ISBN: 354065722-3

[53] Reduced cost of capital has resulted in the out years having real value in discounted cash flow analysis

[54] Jordan DC, Silverman TJ, Wohlgemuth JH, Kurtz SR, Vansant KT. Photovoltaic failure and degradation modes. Progress in Photovoltaics: Research and Applications. 2017;**25**: 318-326

[55] Annigoni E. Reliability of photovoltaic modules: From indoor

testing to long-term performance prediction [PhD thesis]. EPFL; 2018

[56] IEA. Trends 2017 in photovoltaic applications—Survey report of selected IEA countries between 1992 and 2016. Technical Report IEA-PVPST 13-09: 2017. IEA International Energy Agency; 2017

Chapter 5

Reliability Analysis of Photovoltaic Systems for Specific Applications

Laurentiu Fara and Dan Craciunescu

Abstract

This contribution is dedicated to the analysis of a reliable PV system for specific applications. The reliability study was based on: (1) the RAMS (Reliability, Availability, Maintenance, and Safety) model applied to a PV system by using a simulation SYNTHESIS platform developed by ReliaSoft, and (2) the simulation of the PV system using the SYNTHESIS platform and TM-21 Calculator software developed by ENERGY STAR. The objective of this analysis was to obtain a more stable and long-lasting operation of a PV system regarding reliability, maintainability, availability and degradation of the system.

Keywords: reliability, maintainability, availability, degradation, PV system, simulation tools, RBD

1. Introduction

Due to various incentive programs and local market conditions in several European countries, as well as around the world, the PV systems represent a widespread solution for residential houses and other autonomous applications. This approach raises new and important issues related to the efficiency, reliability and safety of the PV systems, either autonomous or integrated in the electrical grid [1, 2].

In the case of grid-connected PV systems, the occurrence of system failures may affect the operation of other interconnected systems, that is why a thorough reliability analysis of such systems should be carried out [3–5]. The reliability analysis also makes it possible to establish an acceptable maintenance plan for both the user and the system as efficiently as possible [6, 7]. However, a reliability analysis is often missing from the feasibility study of photovoltaic systems or is treated superficially due to an incomplete methodological approach or lack of specific simulation tools [8, 9]. A PV system for LED lighting applications has been analyzed [10, 11]. To overcome the reported inconveniences, it is envisaged the reliability analysis, using RAMS (Reliability, Availability, Maintainability, and Safety) statistical models through accelerated testing and application of stress levels [12–15]. Starting from ground studies, the failures of PV systems have shown that manufacturers do not correctly perform reliability analysis of their PV modules' lifetime. For this reason, a small part of the PV modules fails before the lifetime specified by the manufacturer [16]. Reliability analysis applies to both PV systems, and simple systems, such as LED-based lighting lamps [17].

The authors proposed and carried out the reliability analysis of solar cells for specific meteorological conditions [16, 18, 20, 21]. To assess the reliability of the integrated PV system with LED lighting applications, its operating characteristics have been used for a time period [19], and in the case of the simple LED lighting system there were used thermal characteristics of the studied LED lamp [22, 23].

The most relevant analyzes proposed by the authors in this paper are based on reliability block diagrams (RBDs) required for development of RAMS for studied PV systems. RBD diagrams have been developed for: (1) definition of PV systems, (2) description of interdependence between components, (3) identification of how they degrade/fail, and (4) appreciation of how the PV system reliability is influenced.

2. Background, state of the art and simulation tools

2.1 Background

The reliability of a product can be defined as the probability of performing its tasks without incidents within a specified time, under previously established conditions. On this basis, the quality of a product is determined, the reliability concept being applied in almost all areas of engineering, in the preventive maintenance of systems and their components. It is envisaged the modeling of the product's lifetime, represented by the time it has worked successfully or its time until the moment of failure. For the purpose of accurately determining the lifetime, it has been proposed an accelerate product testing (qualitatively or quantitatively during its lifetime) [25, 29].

Qualitative accelerated testing only provides information about system malfunctions or ways of failure. Qualitative tests do not quantify the system's life (or reliability) characteristics under normal conditions of utilization, but provide valuable information on the type and level of demands to use during a later quantitative test. In qualitative accelerated testing, the expected result consists in identifying deficiencies and ways of failure without predicting the life of the product under normal conditions of utilization [2, 24, 26].

Accelerated quantitative testing involves the estimation of product lifetime characteristics, obtained under normal conditions of utilization after an accelerated test. High accelerated life test (HALT) provides various information about the product and its failure mechanisms. Accelerated quantitative testing consists of tests designed to quantify the life characteristics of a product, component or system under normal conditions of utilization, and to provide information on determining the reliability of the system (probability of system failure under various conditions, life expectancy, average lifetime). It can be used to carry out risk assessments or for comparison of different design methods [9].

Two acceleration methods were designed [9] for use of: (1) acceleration rate and (2) stress acceleration, in order to obtain time data on product failure at an accelerated rate. For products where use of acceleration rate is impractical, a second method can be applied, in which the normal stress level of a product is exceeded.

2.2 State of the art

The reliability of photovoltaic systems and their components/elements (solar cells, PV modules, electrical storage systems, inverters, regulators, etc.) are the key issues in manufacturing performance and financially competitive photovoltaic installations. For this purpose, a number of methods for testing and improving the reliability of PV products have been studied [13, 15], such as: (1) the accelerated test method for solar cells [27]; (2) analysis of defect detection and degradation mechanisms detection for solar cells [14], as well as PV systems and their components [28]; (3) energy production forecasting based on the rate of failure and degradation of PV systems and their applications [12].

In this paper the authors proposed to integrate some of these methods by using a specialized complex simulation platform, namely SYNTHESIS, which allows a physical and statistical approach for PV products; it is envisaged the development of a highly useful RAMS methodology, both in research and in the efficient design of PV systems and their applications [29–31].

2.3 Methodology

In order to perform a fairly accurate reliability analysis of the PV systems and their components, the authors of this paper have chosen to use the SYNTHESIS simulation platform developed by **ReliaSoft** [33]. The simulation methodology is intended to explain how the simulation platform modules work. On this basis the obtained results regarding: (1) solar cell, (2) PV system, including its components, and (3) integrated PV system can be understood. The following modules of the simulation platform were used in the analysis:

a. The **Weibull++ software tool** is the industry standard in life data analysis (Weibull analysis) for many companies around the world. This software performs life data analysis using multiple lifetime distributions with a clear and concise, reliability-oriented interface. The features of Weibull++ provide the most comprehensive set of tools available for life data reliability analysis. The software supports all types of data and all distributions commonly used throughout the product's lifetime.

b. The **ALTA software tool** provides an intuitive way of complex and powerful mathematical models for quantitative analysis of accelerated life tests. Accelerated lifetime testing techniques, combined with powerful data analysis methodologies, reduce testing time for the product, thus providing faster product launch times, lower product development costs, as well as lower warranty costs.

c. The **BlockSim software tool** provides a comprehensive platform for the reliability, availability, maintenance and related products analyzes. The software provides a graphical interface that allows modeling of both simple and complex systems and processes using reliability block diagrams (RBDs), failure tree analyzes (FTAs), or combinations of both approaches. Using accurate calculations and/or discrete events simulation, BlockSim facilitates a wide range of analyzes: (1) system reliability analysis, (2) identification of critical components (important for reliability), (3) optimal allocation of reliability, (4) system maintenance analysis (determining the optimal intervals for preventive maintenance), (5) analyzing the availability of the system (determining the operating time, availability/unavailability), (6) performing calculations for obstacle identification, production capacity estimation, etc., (7) estimating the cost of the life cycle.

d. The **RENO software tool** is a platform that allows complex analyzes for any probabilistic or deterministic scenario. Chart models can be created for complex reliability analyzes, risk and safety analyzes, decision making, or maintenance planning. The Monte Carlo simulation methods offer a number of definitions and constructions that allow modeling of the examined situations. Simulation results can be applied to estimate/optimize risk analysis, complex modeling of reliability and maintenance planning.

e. The **RGA software tool** allows the application of reliability enhancement models for data analysis from both development tests and on-site repair systems. In the development phase, the software allows to quantify the increase in reliability achieved for each product prototype and also provides advanced design, planning, and reliability enhancement management methods. The software also offers opportunities for ground-based system analysis, including a test utility for system reliability.

3. Modeling and numerical simulation of PV system components using RAMS analysis: discussion of the results

3.1 Basic concepts

The studied integrated PV system is defined as a PV application, which uses a consumer for a specific application (see **Figure 1**). The implementation of the RAMS analysis of the PV system was performed using the modeling and simulation SYNTHESIS platform for system reliability [32]. When it is analyzed the reliability of studied solar cells, the simulation module used in the SYNTHESIS platform was Weibull/ALTA. This module statistically addresses reliability issues through statistical repartitions and together with the ALTA module contributes to the accelerated life tests. For the analysis of the PV system studied in the paper, the BlockSim simulation module was used within the same platform. The simulation of the PV systems is based on the RBD diagram, as well as on the component types and the way they are arranged in the system.

For a rigorous understanding of the RAMS analysis for the PV integrated system, it was analyzed the highlighting of solar cell degradation. By performing a statistical analysis based on the Weibull, Exponential and Normal repartitions, it was possible to estimate the most frequent events (characterized by main meteorological parameters: solar irradiance, temperature and humidity) that contribute to the most probable degradation of solar cells during their lifetime. The statistical approach was correlated with the accelerating lifetime method by stressing solar cells using different meteorological parameters (**Figure 2**). The present study considered only the three meteorological parameters mentioned above, which have an essential role in cell degradation.

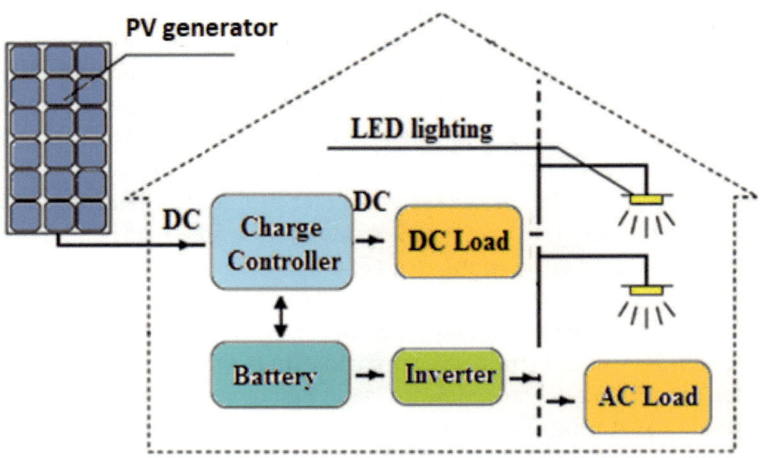

Figure 1.
PV integrated system concept.

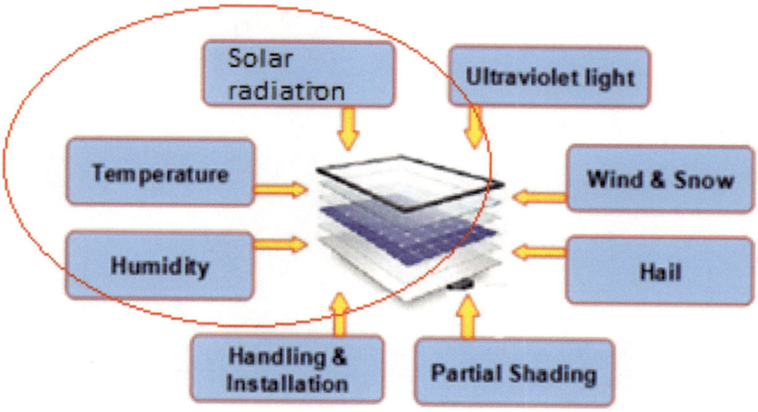

Figure 2.
Accelerated stress test diagram.

3.2 Reliability analysis of solar cells

Reliability analysis of solar cells has been developed based on the accelerated testing method, which has certain advantages: (1) avoiding testing in real and more severe conditions that can lead to cell degradation/failure, (2) determining the frequent reasons that determine cell failure. Highly accelerated stress screening (HASS) aimed at identifying defects in as short time as possible compared to the standard method (based on cell monitoring and results analysis), which may take several months/years. Accelerated testing (HASS) can be performed with either a weather parameter (e.g., temperature/solar irradiance) or combined with few weather parameters (such as, temperature, humidity and solar irradiance). Within this paper the HASS analysis was implemented in the version of the three key meteorological parameters.

The authors' study also looked at establishing the best statistical reparation used to make the most accurate predictions of the of solar cells life time. Three types of statistical repartitions, namely Weibull, Exponential and Normal, were evaluated using the SYNTHESIS software, and based on the correlation cause (determined by one of the three main meteorological parameters)—occurrence (characterized by the degradation degree of the solar cell) it was possible to obtain the probabilities for the three repartitions (the black curve defines the Weibull repartition, the green one defines the normal repartition, and the purple one defines the exponential repartition), using the HASS test. In **Figure 3a** the curves of the mentioned statistical repartitions were obtained. It can be seen that Weibull repartition is the closest to the values of the three main meteorological parameters. The accelerated test has highlighted that the most influential meteorological parameter in solar cell degradation (characterized by most occurrences) is temperature, followed by humidity and further by solar irradiance. **Figure 3b** shows the temporal evolution (in hours) of the degradation level of solar cells for two operating modes: under stress conditions (purple continuous curve) and in standard conditions (blue continuous curve).

Under standard operating conditions, a major degradation level (about 70%) is achieved over a time interval of 2500 h, while under stress conditions the same level of degradation of 70% is achieved over a time interval of 1600 h (see **Figure 3b**).

Using the stress results of the HASS test (purple curve in **Figure 3b**), it was possible to determine by numerical simulation in the SYNTHESIS platform, the limit value of the cell operating time in normal conditions, respectively 1600 h, according to the curve in **Figure 4a**. The accelerated test also enabled the probable number of

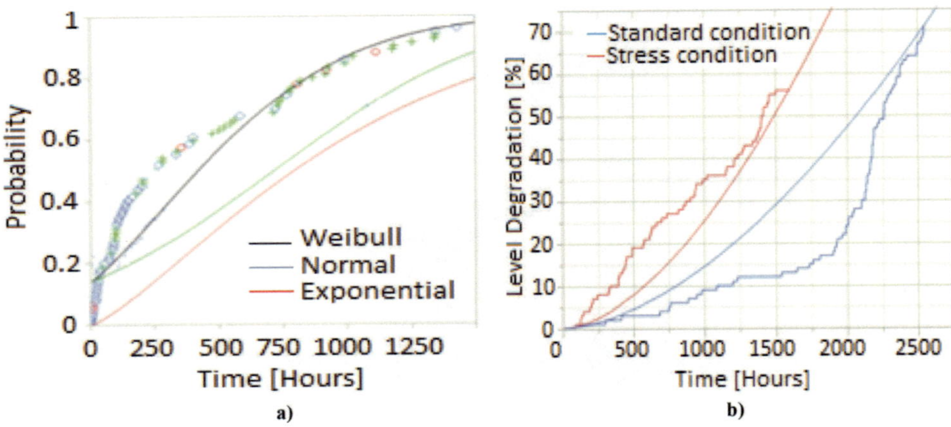

Figure 3.
(a) The life distribution of solar cells for three distribution functions, (b) degradation level of solar cells simulation in two modes: standard conditions and stress conditions.

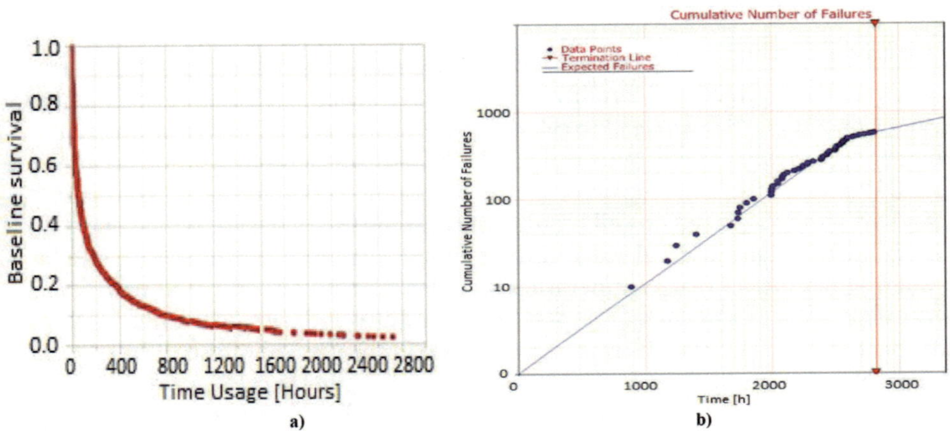

Figure 4.
(a) Baseline survival of solar cells module simulation in stress conditions, (b) cumulative number of failures as function of time.

solar cell failures (see **Figure 4b**) within the range of 1000–2800 h, with a higher concentration between 1500 and 2800 h according to the results from **Figure 4a**; the linear dependence defines the expected failures curve based on Weibull repartition.

3.3 RAMS analysis of the PV system

The analyzed PV system consists of a set of components, arranged to achieve the established performance and to provide acceptable reliability. The reliability characteristics of the PV system and its components can be presented in RBD-type diagrams illustrating their possible failures. These failures are due to a facile analysis of the failures risk.

The RAMS analysis for the photovoltaic system aimed to determine how the system and its components worked over a 5000-h period (about 7 months). For this purpose, it was analyzed the time evolution of the PV generator efficiency, as well as of the PV system's battery efficiency. There were noted high fluctuations for the efficiency of PV generator, slower fluctuations in battery efficiency based on fuzzy logic controller (FLC) and maximum power point tracking (MPPT) methods (see **Figure 5**). The specific application (electric consumer) will be discussed further in Section 3.5.

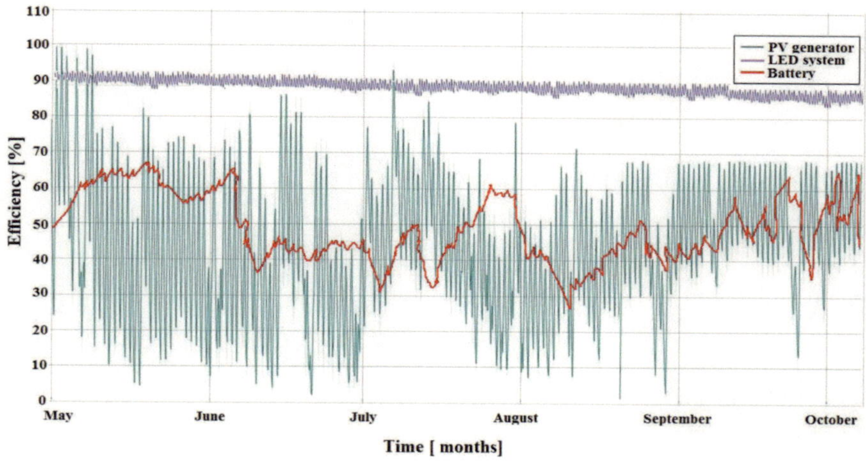

Figure 5.
Efficiency and functionality of a PV system for its main three components: (1) PV generator, (2) battery and (3) electrical load.

Using the results obtained in **Figure 6a** we could establish the availability diagram of the PV system giving the operation time and unavailability periods (when the PV system is not working). The unavailability periods are usually small; after 3000 h of operation there is a 500-h period when the PV system has to be repaired due to different failures. In **Figure 6a** there are also analyzed the main components of the PV system (PV modules, battery and load) to be maintained. The complex analysis of defect risks and preventive planned maintenance of the components aims at avoiding spontaneous malfunctions and premature degradation of the PV system.

In **Figure 6b** the level of degradation for each component of the PV system is presented. Although, excepting the inverter, all other components exceed the critical threshold to the end of the test (3500–4500 hours); however, two components, respectively the PV modules and the storage system, are the most vulnerable to failure (exceeding the critical level much earlier). The simulation tools allowed to obtain the efficiency degradation analysis of the PV system (see **Figure 7a**). A good PV system is characterized by small fluctuations of the time evolution of efficiency, determining a usual degradation based on normal operation requirements for the analyzed simulation period (5000 hours).

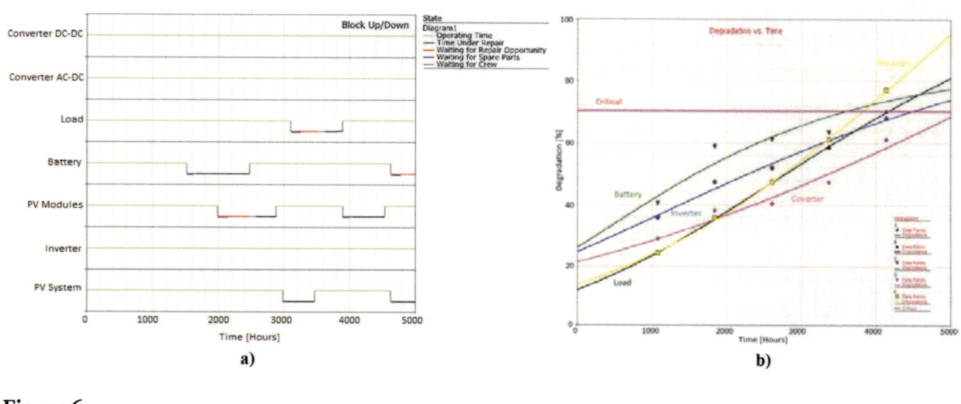

Figure 6.
(a) PV system availability/unavailability diagram, (b) degradation of system components based on availability/unavailability diagram.

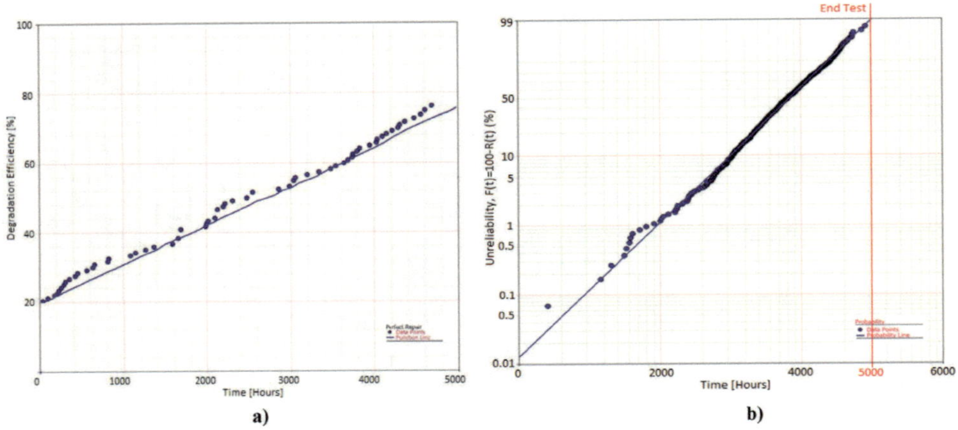

Figure 7.
(a) Time degradation of PV system efficiency, (b) statistical unreliability of the PV system in time.

The probability of non-functioning of the PV system over time is sensitive after 2000 hours of the product use (see **Figure 7b**). It is noted the role played by this diagram in RAMS analysis; when the components of the PV system start to degrade, the number of failures increases, that leads to the need for more frequent interventions (repairs) on the PV system.

The RAMS approach, based on the SYNTHESIS simulation platform, was able to obtain preliminary information on the time behavior of the PV system and its components (from the point of view of its reliability and maintenance) which could be very useful for its proper design.

3.4 RAMS analysis specific to a PV system

A PV system based on specific application has been studied. Analysis of the PV integrated system was possible using the availability/unavailability diagram of its components (**Figure 8**). This diagram describes the interdependence between the

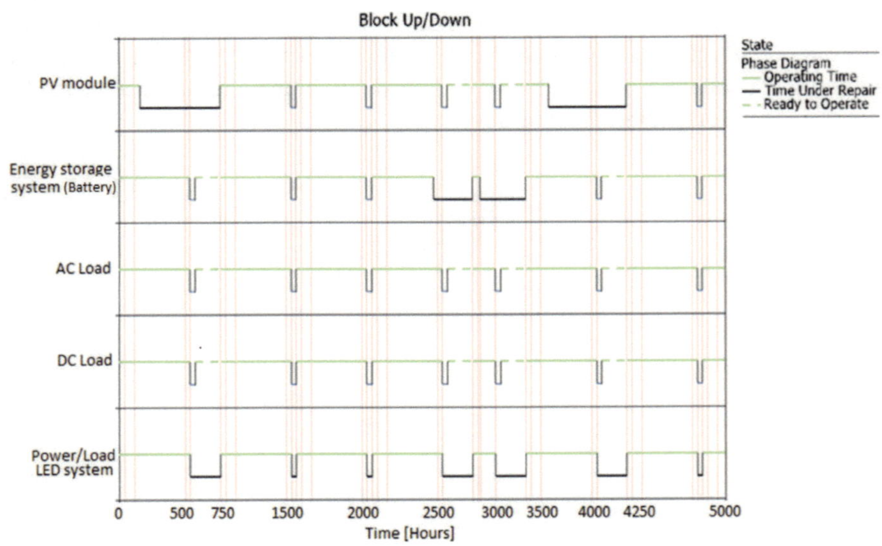

Figure 8.
Diagram of the availability/unavailability (up/down) of the PV integrated system.

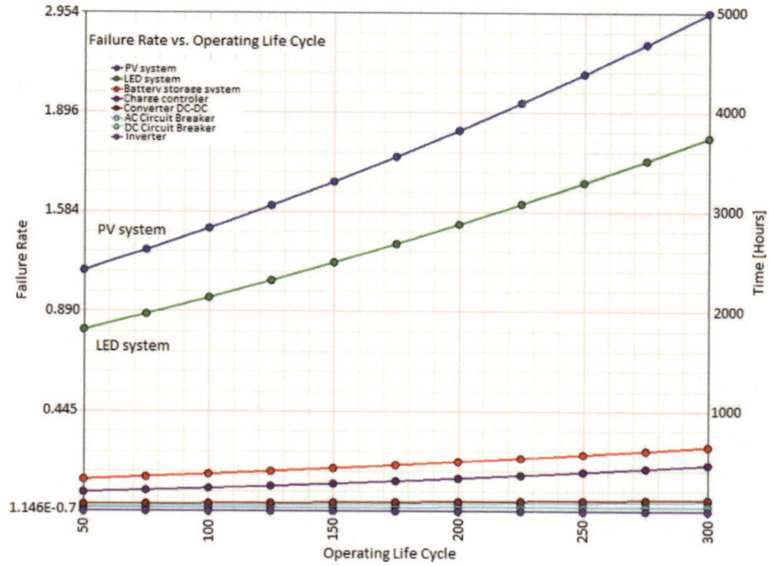

Figure 9.
Time evolution (in hours) of components failure rate of the PV system depending on operating life cycle.

components and defines how the system works. An important parameter of the availability/unavailability diagram is the power load of specific application that is influenced by the operating mode (using either serial, parallel or mixed type configurations). Each component of the PV system is characterized by a repairable block, parameterized in the BlockSim simulation module.

A feature of the up/down diagram is the planning of the PV system maintenance; thus, as a result of prompt interventions in order to repair its components, spontaneous or irreversible failures are eliminated, the system thus becoming functional over the entire lifetime.

The failure rate for each component of the PV system was also being studied. The highest failure rate is presented, by the PV modules, according to the results in **Figure 9**. This is due to the specific configurations for these components (series, parallel or mixed) that increase their failure rate compared with other components (battery, inverter, regulator, etc.).

4. Concluding remarks

The authors have proposed the reliability analysis of a PV system for a specific application using a performance SYNTHESIS simulation platform, which allowed the system to function more accurately by identifying probable defects in certain components of the system.

This contribution is based on an original approach of RAMS reliability analysis. It was considered the adaptation of up/down reliability block diagrams (RBD) from *the reliability theory and applications* for a photovoltaic system in order to determine its functionality and availability and to establish its duration of non-functionality (required to implement the PV system maintenance). The obtained results of this study are in good agreement with the reliability published studies from the point of view of the methods used [1, 2, 6, 7, 17, 26, 28].

The main results obtained by the authors in this article could be synthesized as follows:

- The simulation of the solar cells operation on an accelerated test basis (under the influence of three essential meteorological parameters: solar irradiance, temperature and humidity) allowed to put in evidence the normal operating time under normal/stress conditions, defining the initial degradation time of stressed solar cell (determined by a possible degradation of the cell's I-V characteristics). The Weibull most probable statistical repartition has also been identified using a proper solar cells testing; on this basis, the operating duration in which a significant concentration of defects is achieved, was determined.

- The simulation of the PV system operation allowed to obtain information on the availability/unavailability of the system, based on a RBD diagram. Using this diagram, PV components that may exhibit a higher rate of degradation (PV modules, electrical battery) during operation are identified. Also, useful information on degradation of the PV system was obtained (based on the efficiency degradation curve and the non-functional curve).

The results obtained by the authors were partially validated by other authors research studies, regarding the availability and degradation of analyzed systems [34, 35].

In order to increase the reliability of the PV system and its components, it is necessary to develop an elaborated failure management strategy that involves knowing the initial level of reliability and developing efficient solutions for planning and assessing the probable risks of failure. The essential directions for such a strategy could be as follows:

- experimental accelerated testing of the PV system under the influence of possible degradation factors;

- identifying the cause of failure in experimental conditions, that is intended to isolate possible failures occurring in the PV system;

- implementing effective corrective actions to significantly reduce the failures of the PV system.

Acknowledgements

The authors are grateful for the financial support of Ministry of Research and Innovation, Romanian Executive Agency for Higher Education, Research, Development and Innovation Funding (UEFISCDI), no. 34/2016 contract, SOLHET project "High-performance tandem hetero-junction solar cells for specific applications."

Author details

Laurentiu Fara[1,2*] and Dan Craciunescu[1]

1 Polytechnic University of Bucharest, Bucharest, Romania

2 Academy of Romanian Scientists, Bucharest, Romania

*Address all correspondence to: lfara@renerg.pub.ro

References

[1] Huffman DL, Antelme F. Availability analysis of a solar power system with graceful degradation. In: Proceedings of the Reliability and Maintainability Symposium, Fort Worth, TX; 2009

[2] De Graaff D, Lacerda R, Campeau Z. Degradation mechanisms in Si module technologies observed in the field; their analysis and statistics. In: Presentation at PV Module Reliability Workshop; NREL. Denver, Golden, USA; 2011

[3] Kececioglu DB. Reliability Growth, Reliability Engineering Handbook. 4th ed. Vol. 2. Englewood Cliffs, New Jersey: Prentice-Hall; 1991. pp. 415-418

[4] IEC International Standard. Reliability growth—Statistical test and estimation methods. In: IEC 1164 International Electrotechnical Commission; 1995

[5] Ishii T, Takashima T, Otani K. Long-term performance degradation of various kinds of photovoltaic modules under moderate climatic conditions. Progress in Photovoltaics: Research and Applications. 2011;**19**:170

[6] Crow LH. Evaluating the reliability of repairable systems. In: IEEE Proceedings of the Annual Reliability and Maintainability Symposium; 2004. pp. 73-80

[7] Crow LH. Methods for reducing the cost to maintain a fleet of repairable systems. In: IEEE Proceedings of the Annual Reliability and Maintainability Symposium; 2003. pp. 392-399

[8] Herrmann W, Bogdanski N, Reil F, Kohl M, Weiss K, Assmus M, Heck M. PV module degradation caused by thermo-mechanical stress: Real impacts of outdoor weathering versus accelerated testing in the laboratory. In: Proc. Vol. 7773, Reliability of Photovoltaic Cells, Modules, Components, and Systems III; 777301; 2010

[9] Hoffmann S, Koehl M. Effect of humidity and temperature on the potential-induced degradation. Progress in Photovoltaics: Research and Applications. 2012;**22**:173-179

[10] Sterian A, Sterian P. Mathematical models of dissipative systems in quantum engineering. Journal of Optoelectronics and Advanced Materials. 2012;**2012**. Article ID 347674, 12 pages. DOI: 10.1155/2012/347674

[11] Power Semiconductor Reliability Handbook, Alpha and Omega Semiconductor. 2010

[12] Holley W, Agro S, Galica J, Yorgensen R. UV stability and module testing of non-browning experimental PV encapsulant. In: IEEE Photovoltaic Specialists Conference; 1996

[13] Nelson W. An application of graphical analysis of repair data. Quality and Reliability Engineering International. 1998;**14**:49-52

[14] Kijima M. Some results for repairable systems with general repair. Journal of Applied Probability. 1989;**20**:851-859

[15] Mettas A. Reliability allocation and optimization for complex systems. In: Proceedings of the Annual Reliability & Maintainability Symposium; 2000

[16] Meydbray Y, Wilson K, Brambila E, Terao A, Daroczi S. Solder joint degradation in high efficiency all back contact solar cells. In: IEEE Photovoltaic Specialists Conference; 2008

[17] Collins E, Miller S, Mundt M, Stein J, Sorensen R, Granata J, Quintana M. A reliability and availability sensitivity study of a

large photovoltaic system. In: IEEE Photovoltaic Specialists Conference; 2009

[18] McMahon T. Accelerated testing and failure of thin-film PV modules. Progress in Photovoltaics: Research and Applications. 2004;**12**:235-248

[19] Wohlgemuth J. Tutorial/short course on reliability: PV cells, modules, and systems. In: IEEE Photovoltaic Specialists Conference. Seattle, WA; 2011

[20] Aoki Y, Okamoto M, Masuda A, Doi T. Module performance degradation with rapid thermal-cycling. In: Proceedings of Renewable Energy; 2010

[21] Dhere N, Pethe S, Kaul A. Photovoltaic module reliability studies at the Florida Solar Energy Center. In: IEEE International Reliability Physics Symposium; 2010

[22] Kim SK, Kim SY, Choi YD. In: Proceedings of 10th Intersociety Conference on Thermal and Thermomechanical Phenomena in Electronic Systems; 2006. pp. 377-379

[23] Handbook. Lumileds, understanding power led lifetime analysis. Technical report. Philips Lumileds Lighting Company; 2007

[24] Crow LH. A methodology for managing reliability growth during operational mission profile testing. In: IEEE Proceedings of the Annual Reliability and Maintainability Symposium; 2008. pp. 48-53

[25] Dhere N. Reliability of PV modules and balance-of-system components. In: IEEE Photovoltaic Specialists Conference; 2005

[26] Copper JK, Bruce A. Calculation of PV system degradation rates in a hot dry climate. In: Asia Pacific Solar Research Conference; Canberra, Australia; 2016

[27] Desombre A. Methodology for a reliability study on photovoltaic modules. In: Proceedings of the third European Commission, PV Solar Energy Conference. Cannes, France; 1980. pp. 741-745

[28] Vazquez M, Rey-Stolle I. Photovoltaic module reliability model based on field degradation studies. Progress in Photovoltaics. 2008;**16**:419

[29] Wohlgemuth J. Long-term photovoltaic module reliability. In: NCPV and Solar Program Review Meeting, NREL/CD-520-33586; 2003. pp. 179-183

[30] Van Weeran N. Degradation of photovoltaic modules in Alice Springs, Australia [undergraduate thesis]. School of Photovoltaics and Renewable Energy Engineering. The University of New South Wales; 2013

[31] Granata J, Boyson W, Kratochvil J, Quintana M. Long-term performance and reliability assessment of 8 PV arrays. In: IEEE Photovoltaic Specialists Conference; 2009

[32] Craciunescu D, Fara L, Sterian P, Bobei A, Dragan F. Optimized management for photovoltaic applications based on LEDs by fuzzy logic control and maximum power point tracking. Nearly Zero Energy Communities. Springer Proceedings in Energy. 2017. DOI: 10.1007/978-3-319-63215-5_23

[33] Synthesis Platform. Available from: http://www.reliasoft.com/synthesis/downloads.htm [Accessed October 2017]

[34] Mahdi I, Chalah S, Nadji B. Reliability study of a system dedicated to renewable energies by using stochastic petri nets: application to photovoltaic (PV) system. In: 4th International Conference on Energy and Environment Research, ICEER 2017.

17-20 July 2017; Porto, Portugal. Energy Procedia 136; 2017. pp. 513-520

[35] Lavrova O, Flicker J, Johnson J, Armijo K, Gonzalez S, Schindelholtz E, Sorensen R, Yang B. PV Systems Reliability: Final Technical Report. Sandia Report-SAND2015-376607; 2015

Chapter 6

Emerging Thin Film Solar Panels

Mahmoud Zendehdel, Narges Yaghoobi Nia and Mohammadreza Yaghoubinia

Abstract

Utilizing of photovoltaics (PVs) has been rapidly developing over the past two decades due to its potential for transition from fossil fuels to renewable energy based economies. However, PVs as fuel less energy sources will be sustainable if some issues such as raw materials abundance, production cost, and environmental impacts carefully addressed in their value chains. Among PV technologies, thin film solar panels have been illustrated the potential to reach the sustainability. In this chapter we review some studies about environmental impacts of thin film PVs through life cycle assessment (LCA) and some environmental fate modeling. For the PV technologies, LCA studies need to be conducted to address environmental and energy impacts and encourage the development of PV technologies in a better sustainable way. Three methods of impact assessment in LCA are reviewed and compared, namely, Energy Payback Time (EPBT), Cumulative Energy Demand (CED), and Greenhouse Gases (GHG) emission rate, owing to data and information published in the literature. Generally, most results show promising potential of emerging thin film PVs, especially perovskite solar cells, to reach the best sustainable solution among PV technologies in near future.

Keywords: perovskite solar cell, life cycle assessment, emerging thin film photovoltaics, environmental impact, sustainable energy

1. Introduction

1.1 Sustainability of solar energy

The most requiring environmental concerns which need to be addressed are impacts of fossil fuels, Climate change, resource depletion, and worldwide energy shortage. The mentioned concerns will be more fundamental when we want to find the sustainable energy solutions for the future. Solar energy as the most abundant natural power resource on earth, can generate renewable energy by converting sunlight to thermal or electrical power by using of the photovoltaic (PV) devices. The effective solar irradiance which arrive the earth's surface varies between 125 and 305 Wm^{-2} by considering the latitude-dependent oblique incidence, diurnal variation, and seasonal variation. As it is known, solar energy is one of the continual power sources that could provide energy independence and energy security for all nations. At the moment, solar power is adopted as a substantial electricity generation in many developed and developing countries in order to address the energy demands. However, PVs as fuel-free energy sources inherently will be sustainable unless they are too expensive to produce, the materials needed for their production are depletable, or they create serious environmental impacts.

Net annual CO_2 emission mitigation potential from 1.8 kWp solar PV pump at an average solar radiation of 5.5 $kWhm^{-2}$ is about 2085 kg from diesel pumps and about 1860 kg from petrol pumps. An investigation on generation costs and carbon emissions of the conventional energy production technologies compared with solar power generation is presented in **Table 1** [1].

1.2 Photovoltaic technologies

The present PV technologies could be classified in two categories: (1) wafer-based (2) thin film cells (**Figure 1**). Wafer-based cells are fabricate on semi-conducting wafers and could be handled without an additional substrate, while modules are typically covered with glass for improving the mechanical stability and more protection. Thin film cells consist of semiconducting layers which can be deposited onto various substrates such as plastic, glass or metal. Further division of thin films into commercial and emerging thin film technologies is presented in **Figure 1**.

1.2.1 Wafer-based technologies

Three primary wafer-based technologies exist in the market containing: (1) Crystalline Silicon (c-Si), (2) Gallium Arsenide (GaAs) and (3) III-V multijunction (MJ). Among these types, c-Si PVs have occupied ~90% of present global manufacturing capacity and are the most mature of all PV technologies. Silicon solar cells are divided as single-crystalline (sc-Si) or multicrystalline (mc-Si). The sc-Si is typically fabricate through Czochralski (CZ) [3] or float-zone (FZ) methods, while mc-Si ingots are produced by casting technique. Usually, the fabricated ingots are sliced into 150–180 μm wafers prior to cell processing. High photoconversion efficiencies of sc-Si can be achieved by heterojunction with intrinsic thin layer (HIT) architecture, which consists of n-type sc-Si with thin amorphous silicon films. The mc-Si cells contain randomly oriented silicon grains with average size of 1 cm^2. Present record cell efficiencies stand at 27.6% for sc-Si and 22.3% for mc-Si [4]. Indirect bandgap of c-Si is one fundamental limitation, which leads to weak light absorption and needs wafer structures with ~100 μm thicknesses, excluding advanced light-trapping strategies. Key technological challenges for c-Si consist of stringent material purity requirements, large volume material utilization, restricted module form factor, and batch-based cell fabrication and module integration processes with relatively low throughput.

The GaAs is well suited for solar energy conversion, because of, a direct bandgap well matched to the solar spectrum, strong absorption and very low non-radiative energy loss. For lab cells the highest power conversion efficiencies of 30.5% has achieved by GaAs [4]. Flexible thin films of GaAs can be fabricated through epitaxial liftoff technique which can amortizes the substrate costs by recycling of GaAs wafers [5]. However this method has not yet been developed in high production scale. There are still some economical aspects such as, need to improve the film quality, more substrate recycling, and low-cost polishing of wafer, which need to be

Technology	Carbon emissions (gC/kWh)	Generation costs (USD/kWh)
Solar thermal and solar PV systems	0	9–40
Pulverized coal-natural gas turbine	100–230	5–7

Table 1.
Economic and emissions of conventional energy production technologies compared with solar power generation.

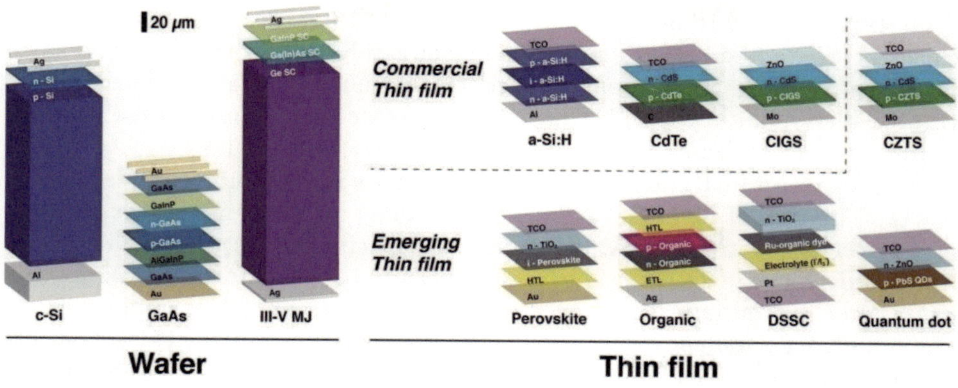

Figure 1.
Typical PV device structures, divided into thin film and wafer-based technologies. Primary absorber layers are labeled in white color, and thicknesses are shown to scale (obtained under copyright of Royal Society of Chemistry [2]).

addressed. In any case, high precursor costs may limit the large-scale development of such PV systems.

The III-V multijunction (MJ) solar cells stack of two or more single-junction cells with different band gaps which are used for absorbing of the solar irradiation with minimum thermalization losses. As it is known, semiconducting compounds of group V (N, P, As, Sb) and group III (Al, Ga, In) elements can form crystalline films with variable band energies, yielding unparalleled power conversion efficiencies, 35.5, 44.4, and 46.0% for record 4-junction (2 J), 3 J, and 4 J cells, respectively, under concentrated illumination [4]. One of the leading technologies for space application is III-V MJs, owing to their high radiation resistance, high efficiency and low temperature sensitivity. However, complex manufacturing processes in addition to high material costs make III-V MJ cells very expensive for large-area terrestrial applications.

1.2.2 Commercial thin-film PV

As it can be seen that c-Si currently dominates the global PV market, but alternative technologies may be able to reach lower costs in the long run. Solar cells based on thin semiconducting films constitute ~10% of global PV module market nowadays. Generally, thin-film cells are made by additive fabrication processes, which may reduce manufacturing capital expense and material usage. This category extends from commercial technologies based on conventional inorganic semiconductors to emerging technologies based on nanostructured materials. At the moment (2019), three thin film PV technologies which are developed to commercial phase are (1) hydrogenated amorphous silicon (a-Si:H), (2) cadmium telluride (CdTe) (3) copper indium gallium diselenide ($CuIn_xGa_{1-x}Se_2$, or CIGS).

Hydrogenated amorphous silicon (a-Si:H) offers higher absorption compare to c-Si, although having larger band gap respect to the c-Si (1.7–1.8 eV, compared to 1.12 eV for c-Si) is not well matched to the solar spectrum [6]. Amorphous silicon as a thin film PV is typically fabricated by plasma-enhanced chemical vapor deposition (PECVD) at relatively low substrate temperatures of 150–300°C. Since, only a 300 nm film of a-Si:H can absorb ~90% of above band gap photons in a single pass, this properties led to having lightweight and flexible solar cells and panels. Another properties of a-Si:H cell is that it can be combined with cells based on nanocrystalline silicon (nc-Si) or amorphous silicon-germanium (a-SiGe) alloys to form a multijunction cell without lattice-matching requirements. Nowadays most commercial a-Si:H modules use multijunction approach. As we know silicon is cheap, non-toxic and earth abundant, but while a-Si:H cells are well suited for small scale and low-power applications, their properties such as light induced degradation (the Staebler-Wronski effect [7]) and low efficiency compared to mature thin film technologies led to decreasing market interest.

Leading thin-film PV in the present global market is Cadmium telluride (CdTe). This type of PV is a favorable semiconductor for solar energy harvesting, via a direct band gap of 1.45 eV and strong solar spectrum absorption. CdTe has record efficiencies of 22.1% for the lab-scale cells and efficiencies of the commercial module continue to improve steadily [4]. CdTe technologies employ high throughput deposition processes and the lowest module costs of any PV technology on the market offered by CdTe technology, although relatively high processing temperatures are required (~500°C). However, one of the main concerns that have motivated research on alternative material systems is about the toxicity of elemental cadmium and the scarcity of tellurium.

Copper indium gallium diselenide (CIGS) is a semiconductor composite with a direct band gap of 1.1–1.2 eV. This class of thin film PVs can be fabricated by a variety of solution- and vapor-phase techniques from polyimide substrates or flexible metals [8], which make CIGS as a favorable PV for building-integrated and other unconventional PV applications. CIGS solar cells exhibit high radiation resistance which is mandatory for space applications. For the concentrator cells record efficiencies stand at 23.3% [4]. Some of Key technological challenges are listed as: (1) high variability in film stoichiometry and physical properties, (2) limited knowledge of the grain boundaries activity [9], (3) low open-circuit voltage due to structural and electronic inhomogeneity [10], (4) engineering of higher-band gap alloys to enable multijunction devices [11]. One reason could hinder large-scale deployment of CIGS technologies is the scarcity of indium element. The active materials which are used as light absorber in commercial thin-film PV technologies can absorb the sun light 10–100 times more efficiently than silicon, allowing use of films just a few microns thick. Low precursor materials use is thus a key advantage of these commercial thin film technologies which can affect on the LCA results as well.

1.2.3 Emerging thin-film PVs

Recently, several new thin-film PV technologies have emerged as a result of intense R&D efforts in materials discovery and device engineering. Key emerging thin-film PV technologies could be classified to 5 technologies consist of, (1) copper zinc tin sulfide (Cu_2ZnSnS_4, or CZTS), (2) perovskite solar cells (PSCs), (3) organic photovoltaics (OPV), (4) dye-sensitized solar cells (DSCs) (5) colloidal quantum dot photovoltaics (QDPV).

CZTS is an Earth-abundant alternative technology respect to CIGS, with similar processing challenges and strategies [12]. One of the important challenges consists of uncontrolled Cu and Zn inter-substitution which led to point defects, blocking charge extraction and reduce the open-circuit voltage of the cell [13]. Record certified cell efficiencies of CZTS cells have gained 12.6% [4].

Perovskite solar cells (PSCs) have been raised from solid-state dye-sensitized solar cells [14] and have quickly illustrated as one of the most promising emerging thin-film PV technologies, achieving the certified efficiencies of 24.2% [4] in 3 years of development in lab-scale devices. The term “perovskite” refers to the ABX3 crystal structure, and the most widely investigated perovskite for solar cells is the hybrid organic–inorganic lead halide (MA and/or FA)$Pb(I, Cl, Br)_3$. One of the interesting properties of perovskite structures is high band gap tunability in the range of 1.25–3 eV by substitution of cation or anion in the lattice (e.g., $HC(NH_2)_2Pb(I_{1-x}Br_x)_3$ [15], $CH_3NH_3SnI_3$ [16], and $CH_3NH_3Pb(I_{1-x}Br_x)_3$ [17]). Various low temperature techniques for the solution or vapor deposition can be successively utilized for fabrication the perovskite thin films [18, 19]. The important structural/physical properties of such hybrid perovskite layers could be classified as, long diffusion lengths of charge carriers [20], low recombination reaction [21], low cost precursor materials, and wide band gap tunability. Open circuit voltage (Voc) is one the most difficult parameters to improve in PV devices, while recently PSCs could achieved high Voc (more than 1.1 V). However, there are many issues which should be considered such as device life span, high sensitivity to the water, control the film morphology and reproducibility from batch to batch of fabrication, and present of toxic lead element.

As it is known in OPV generally organic small molecules [22] or polymers [23] are used for absorb the light. These materials have some specific properties such as using of earth abundant materials and easy scale-up the thin film structures by using of various deposition techniques [24]. Furthermore, fabrication of organic multijunction devices could be more economic in compare to the conventional III-V MJs due to high defect tolerance and leisure deposition routes [25]. At the moment the lab-made OPV cells could reach to 15.6% certified efficiencies [4]. Key concerns of these systems involve inefficient exciton transport [26], poor long-term stability [27], low large-area deposition yield, and low ultimate efficiency limits [28].

DSCs are among the most mature of nanomaterial-based PV technologies [29]. These photoelectrochemical cells consist of a transparent inorganic scaffold anode (typically nanoporous TiO_2 or other n-type oxides of transition metals) sensitized with light-absorbing dye molecules (ruthenium (Ru) complexes or organic dyes). Unlike the other solid-state technologies which are discussed here, DSCs often utilize a liquid electrolyte for transport ions to a counter electrode. DSCs have achieved efficiencies of up to 12.3% [30] and may benefit from low-cost materials, colorful and flexible modules and simple assembly. Main challenges of DSCs involve limited long-term stability under illumination and high temperatures, low absorption in the near-infrared, and low open-circuit voltages created from interfacial recombination.

QDPV technologies are improving consistently, with a record certified cell efficiency of 16.6% [4]. QDPV, also known as quantum dots (QDs), use solution-processed nanocrystals for absorb the sun light [31]. The ability to tune the band gap of colloidal metal chalcogenide nanocrystals (primarily PbS) by changing their size allows efficient harvesting of near-infrared photons, as well as the potential for multijunction cells using a single material system [32]. QDPVs have some advantages such as simple room-temperature fabrication and air-stable operation [33]. However, incomplete knowledge of surface chemistry and low open-circuit voltages that may be limited fundamentally by mid-gap states or inherent disorder in QD films make some challenges though development of these class of emerging PVs [34].

These emerging thin-film technologies employ nanostructured materials that can be engineered to achieve desired electronic and optical properties. Earth-abundant materials and relatively simple processing methods open a promising gate for large-scale manufacturing and deployment of such emerging PVs. However despite the unique device properties (e.g., transparency, light weight and flexibility) of these technologies, the maturity from R&D to industrial production and commercialization has not yet been developed. In any case, we believe that the mentioned emerging thin-films can solve many problems of present PV market due to their specific properties and ultra low production cost.

1.3 Life cycle assessment

Life Cycle Assessment (LCA) is a universal model for study the impacts of a system during its life cycle and production chain. This model is very useful method for evaluation of the environmental impacts, energy consumption and economical aspects of a system or product. In the LCA usually a "cradle-to-grave" or "cradle-to-gate" rough has been utilized for different steps in life cycle such as, raw material extraction, transport, processing, fabrication, distribution, utilization, and disposal aspects [35]. Owing to the standard protocol of ISO 14040 [36] and ISO 14044 [37], the LCA methodology consists of four distinct steps as shown in **Figure 2**.

1. Life Cycle Inventory (LCI): consist of data collection from the system (e.g., precursor materials, amounts of energy and water consumption, fabrication routes, waste treatment, products and environmental emission of by-products).

2. Life Cycle Impact Assessment (LCIA): Explain the potential environmental impacts, (e.g., eco-toxicity, ozone depletion, global warming and acidification).

3. Goal and scope definition: Determine the process roadmap, set the system's boundaries, define the functional unit, and formulate some assumptions.

4. Interpretation: conclude the LCI and LCIA results, analyze the critical points, and make recommendations for future improvements.

An effective LCA will quantify the technological, environmental, economic, and social aspects of a product/process that plays a fundamental role for reducing the life cycle impacts toward more sustainable options.

According to the Methodology Guidelines for the LCA of PV systems [38], the life cycle of a PV technology or product starts from the extraction of raw materials and ends with the disposal or recycling, as illustrated in **Figure 3**. During the

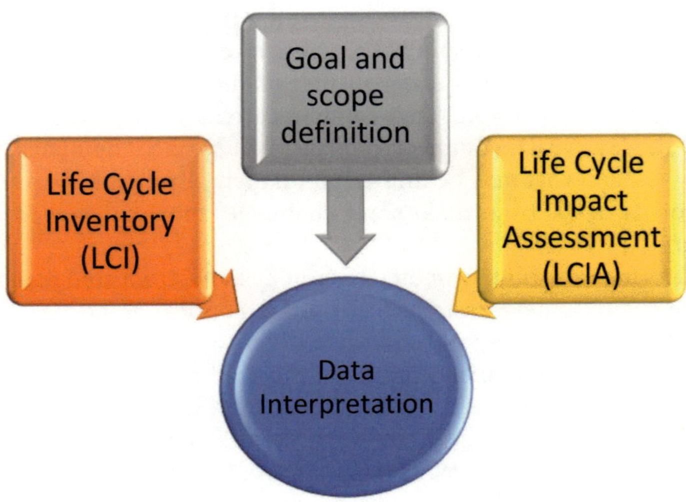

Figure 2.
General framework of life cycle assessment (LCA).

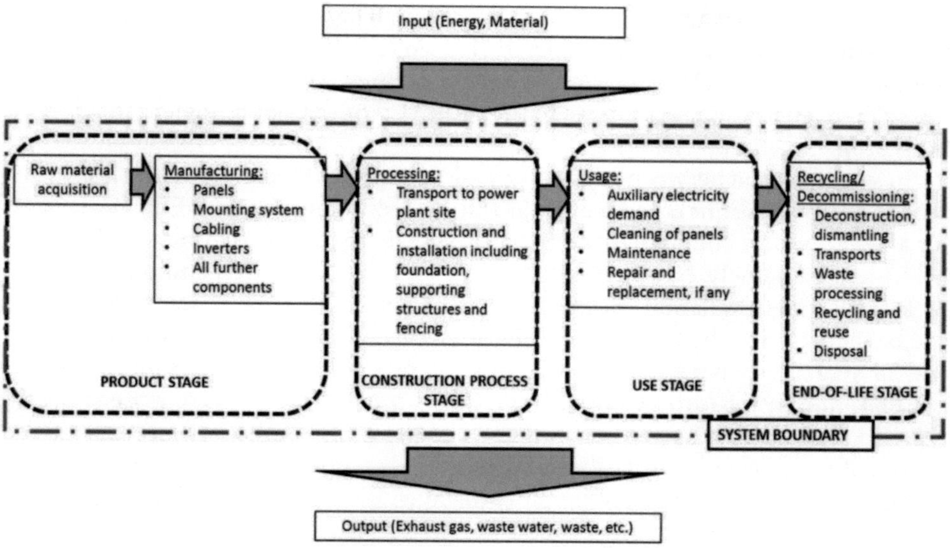

Figure 3.
The system boundary of solar PV life cycle (the figure is obtained under copyright of Elsevier [35]).

production section, insertion of raw materials, energy supply, manufacturing of panels, mounting system, cables, inverters, and all other components needed to produce electricity from sun light should be included in the system boundaries. Furthermore, some activities such as transportation, construction, and installation of the products, should be evaluated in the construction stage. In the utilization stage, some aspects, such as auxiliary electricity request, cleaning of the panels and maintenance of the plant should be included in the system boundaries. Finally, the recycling issues and waste treatments should be studied in the end-of-life stage.

Accordingly, the goal of this LCA is analysis the environmental impacts and other problems of PV systems during their life cycle which make a useful vehicle for improvement of each technology. Generally, this evaluation realizes three indicatives solar PV performance containing, Cumulative Energy Demand (CED), Energy Payback Time (EPBT), and GHG emission rate or Global Warming Potential (GWP). Those parameters are the most usable metrics applied in comparative life cycle evaluations of PV system.

The Cumulative Energy Demand (CED) is an energetic factor that quantifies the energy requirement during the life cycle of a product.

The Global Warming Potential (GWP) method, developed by the International Panel of Climate Change (IPCC), assesses the impact of the whole life cycle on global warming issue in terms of carbon dioxide (CO_2) emission.

The EPBT is an energetic indicator usually used for comparison of different energy production technologies, like PVs. It is represented in years and shows the needed time for generate the same amount of system's consumed energy in life cycle.

2. Life cycle assessment of thin film PVs

2.1 LCA of commercial thin film PVs

Environmental impacts of solar PV systems have been illuminated by several recent researches. Despite huge manufacturing of the crystalline silicon based PV systems, these generations of PVs have been shown higher environmental impact in comparison to thin film PVs. In particular, CdTe PV systems by presenting low life cycle impacts and low production costs compared with c-Si based PVs, can be considered as sustainable solution to the future Cd oversupply problem that has been projected in East Asia [39].

In a comparison between crystalline and amorphous silicon PVs (first and second PV generations), Pacca et al. [40] have been shown shorter EPBT values for a-Si (3.2 years) compare to mc-Si (7.5 years). In this study they assessed three LCA metrics, namely, Net Energy Ratio (NER), EPBT, and GHG emissions, on a 33-kW rooftop installation. However, if the photoconversion efficiency of cells can be increased in the future, the EPBT for a-Si and multi-Si could drop to 1.6 and 5.7 years, respectively. The GHG emission rate also exhibits the same pattern as EPBT, presenting lower values (34.3 g $CO_{2\text{-eq}}$/kWh) for a-Si in compare to mc-Si (72.4 g $CO_{2\text{-eq}}$/kWh). In another work, Minemoto et al. [41] have been reported the impact of solar spectrum distribution and module temperature on the outdoor performance of a-Si and mc-Si PV modules installed. Owing to their results, the output energy of a-Si modules mainly depends on spectrum distribution and is higher under blue-rich part of spectrum while that is less sensitive to module temperature.

The stages of the CdTe PV life cycle during thin film manufacturing has been reported by Fthenakis [42]. Firstly, Cd was extracted from zinc ores (~ 80%), while Te was prepared from Cu ores. Cd and Te were subsequently purified to more than 99.99% through electrolytic purification. Subsequently, a transparent conducting oxide (TCO) layer was deposited onto a glass substrate and a thin film of CdS followed by a CdTe layer was deposited via vapor deposition, followed by spray coating/thermal treatment of $CdCl_2$. Finally, the CdTe solar cell was completely fabricated by sputtering of a metal layer as back contact. Owing to fabrication of CdTe module, the individual cells were interconnected in series using laser ablation, followed by lamination, in which glass plates were placed and thermally sealed with the glass substrate. At the end of process, the module was encapsulated between two glass plates to form the final module with a less than 10 μm film thickness.

In another work, LCA analysis has been run for CdS/CdTe PV modules in order to estimate primary energy demand, EPBT, and GHG [43]. The results showed EPBT values of 1.7, 1.4 and 1.1 years for the plant scales of 10, 30 and 100 MW/year, which are significantly lower than the EPBTs of a-Si and mc-Si results in the same plant scale. The GHG values of CdS/CdTe PV modules have been ranged as 8.9–14.0 g $CO_{2\text{-eq}}$/kWh for 10–100 MW/year productions. Ito et al. [44] have been illustrated a comparative LCA for a 100-MW PV ground mounting plant, using mc-Si, a-Si, CdTe, and CIS at different module efficiencies. Owing to their results, CIS modules have shorter EPBTs of 1.6 years compared with those of CdTe and a-Si modules, which have 1.9 years and 2.5 years, respectively. The GHG value for these modules has been evaluated in the range of 9–16 g $CO_{2\text{-eq}}$/kWh.

In particular, Fthenakis and Kim [45] conducted LCA studies on thin film technologies, especially CdTe PV modules. With particular discuss about cadmium emissions during the life cycle of modules. Indeed, Zinc smelting/refining is most atmospheric Cd emitting stage with 40 g Cd/ton, They also evaluated the GHG and heavy metal emissions from c-Si, mc-Si, ribbon-Si, and thin film CdTe modules [46] (**Figure 4a**). Environmental emission of cadmium metal from CdTe modules compared with other sources of electricity which illustrate significantly lower heavy metal impact compare to the life cycle of common fossil fuels (see **Figure 4b**). These environmental cadmium emission results show the importance of a general LCA analysis to make a true decision for national renewable energy investment. While CdTe PV systems use Cd element as a main absorber component,

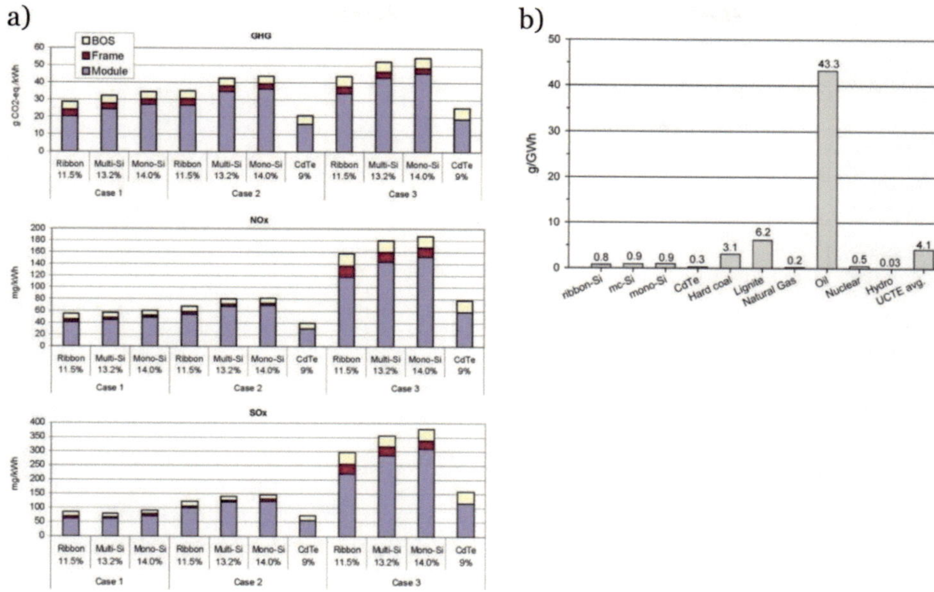

Figure 4.
(a) Life-cycle emissions from silicon and CdTe PV modules. BOS is the balance of system. (b) Life-cycle atmospheric Cd emissions for PV systems from electricity and fuel consumption. The figures are obtained under copyright of American Chemical Society [46].

the emission values show just 0.3 g/GWh which is comparable with the natural gas sources and lower than typical silicon PVs.

In another research, Rocchetti and Beolchini used LCA in order to the management of end-of-life CIGS and CdTe panels [47]. They compared the environmental impacts of the recycling processes for such PV panels, compare with landfill site disposal. Two recycling situation have been evaluated containing conventional and innovative. By using of the conventional method, the modules are crushed and glass components are mechanically recovered. The ethylene-vinyl acetate (EVA) sealing material is passed to thermal treatment for energy generation and the residual material is disposed in a landfill site. On the other hand, the innovative process leads to recover also selenium, indium and gallium from the CIGS panels and tellurium from the CdTe panels as well. However, the potential environmental impacts were similar for the conventional recycling process of both PV panels. Conversely, the innovative recycling of the CdTe panels creates a net production of environmental credits thanks to the recovery of valuable materials. The innovative recycling of CIGS panels has been shown a higher impact compare to the recycling of CdTe panels (2.5 vs. 0.7 kg $CO_{2\text{-eq}}$., respectively, for GHG). In any case, according to LCA results, the disposal of end-of-life panels is not advantageous for the environment. Data obtained with the recycling processes suggest that the innovative recycling is environmentally beneficial only for the CdTe panels, due to the very low content of valuable elements in the CIGS panels. The potential impacts of the management options for the CIGS and CdTe panels in several categories are presented in **Figure 5**. Due to the environmental loads problems, the innovative recycling approach for the CIGS panels was not favorable, in which this refinement option had a high demand of raw materials and energy. In particular, the mentioned recycling management shows a positive correlation with the GHG emission, which generally leads the impact assessment. However, in all categories it is clear that disposal in landfill sites was the worst scenario for the environment.

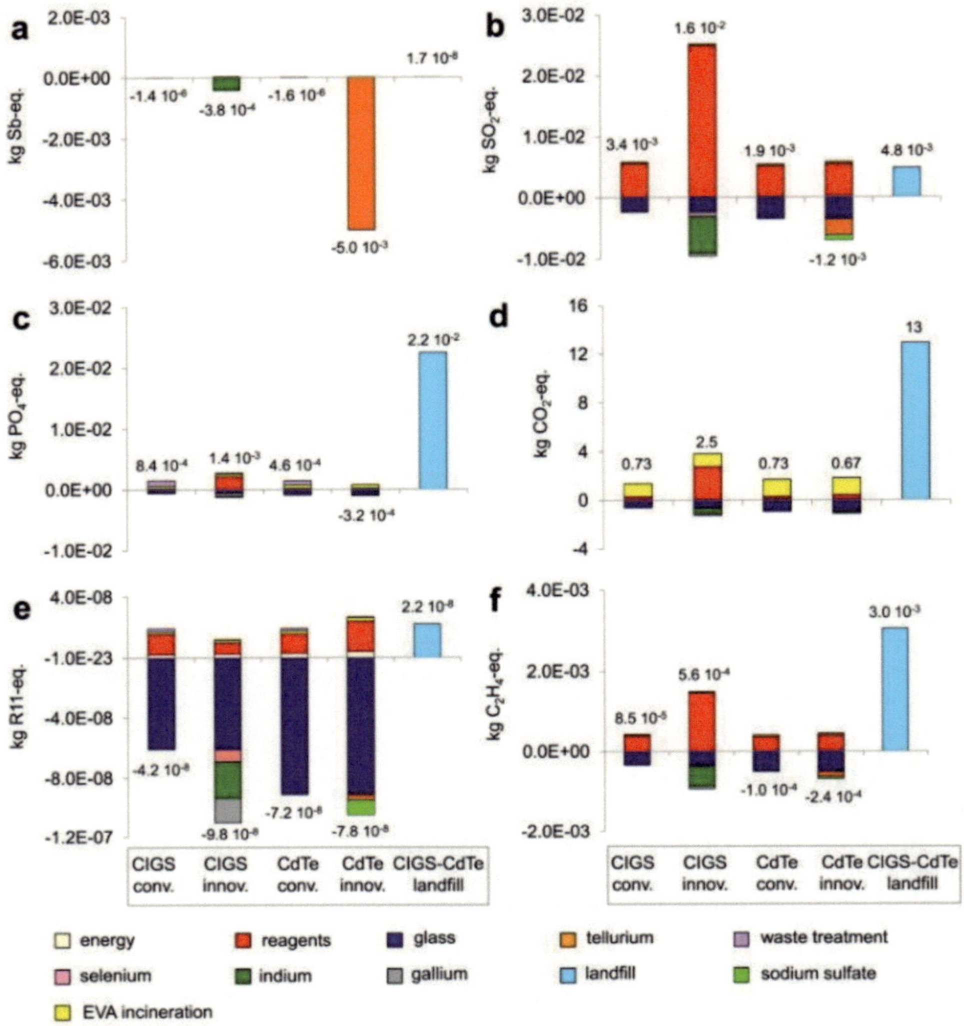

Figure 5.
Potential impacts of the management options for the CIGS and CdTe panels in the categories: (a) abiotic depletion; (b) acidification; (c) eutrophication; (d) global warming; (e) ozone layer depletion (f) photochemical ozone creation. The figure is obtained under copyright of Elsevier [47].

Waste PV panels, depending on the types of PV cells, have different environmental impact potentials due to different contents of substances. Bang et al. [48] have been compared hazardous waste and toxicity potentials, resource depletion from metals in three types of PV modules (CIGS a-Si, mc-Si). Resource depletion and toxicity potentials have been evaluated by using life cycle impact assessment methods and Hazardous waste potentials were examined by using metal leachability tests (**Figure 6**). It is interesting that a-Si PV does not hazardous waste due to Pb/Cadmium or selenium while mc-Si and CIGS PVs have hazardous waste potentials due to lead (Pb) and cadmium/selenium, respectively. Regarding the resource, the mc-Si shows the highest depletion potential due to silver; next highest depletion potential is related to the CIGS due to selenium element; while the lowest rate is for Si PV which is depended to tin and copper. For toxicity potentials, overall the a-Si had lower potentials, derived primarily from Ba/Cu/Ni/Zn elements, than the mc-Si and CIGS PVs of which the toxicity potentials were primarily form Cu/Pb/Ni/Ag and Cu/Hg/Mo/Ni/Ag, respectively. Indeed, waste mc-Si and CIGS PV panels should be recycled and managed with priority.

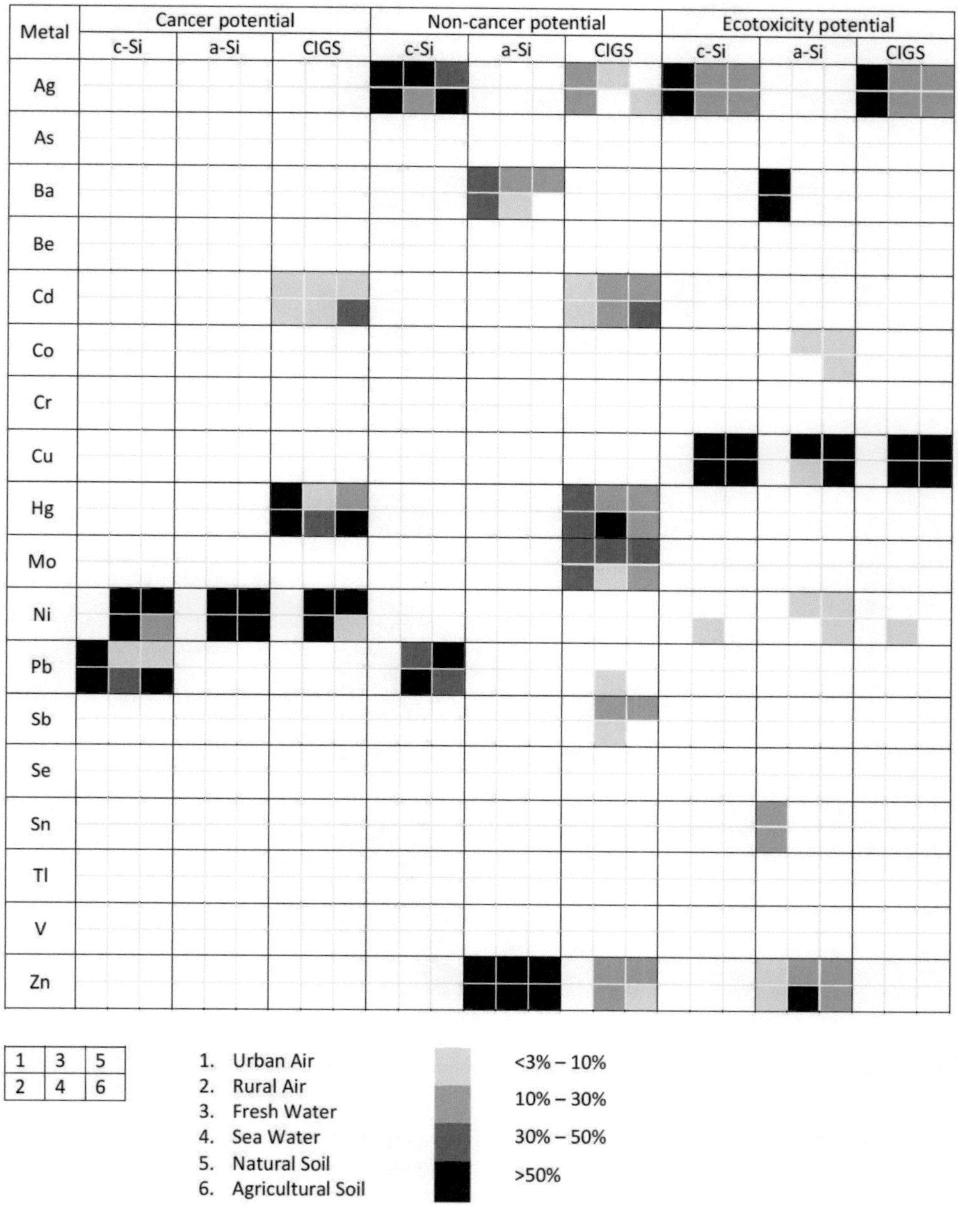

Figure 6.
Toxicity potentials from metals in the PV modules. The share means the proportion of the potential from a given metal to the total from all the metals in each module [48].

2.2 LCA of emerging thin film PVs

Dye-sensitized solar cells (DSCs) are a class of nanomaterial based PVs which show some benefits among other commercial PV systems due to lower production cost and environmental impacts. In the typical DSC fabrication, firstly a transparent conducting oxide (TCO) (e.g., fluorine-doped tin oxide (FTO) and indium-doped tin oxide (ITO)) has been deposited on glass substrates and then it has been covered with a semiconductor material layer, which act as scaffold layer of photo-anode. Various nanomaterials and nanocomposite structures have been applied as scaffold photoanode in DSCs [49–55]. After deposition, the coated photoelectrode is sintered in the furnace. Dye sensitizer molecules, such as N719 or organic dyes

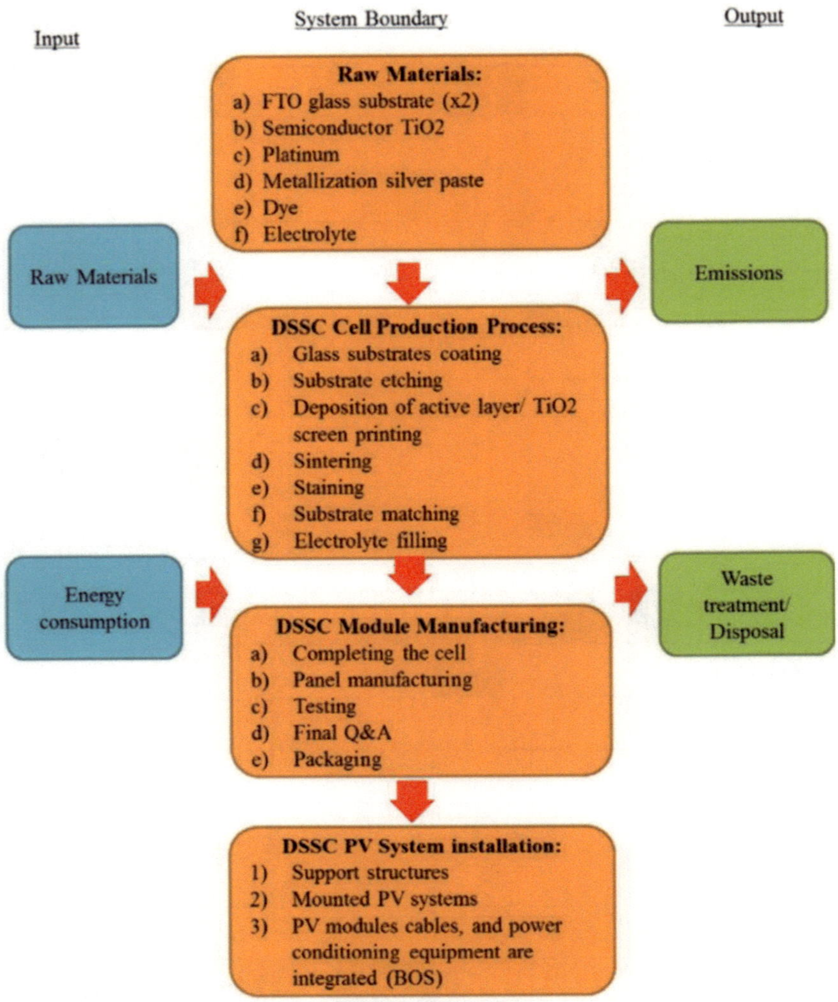

Figure 7.
LCA for DSC process from cradle-to-gate. The figure is obtained under copyright of Elsevier [35].

(such as D35), are then anchored on the photoanode surface to harvest and enhance light absorption [56]. The dye is surrounded by electrolyte solution, containing I^-/I_3^- [57] or Co(II)/Co(III)-complex [58, 59] redox mediators which are dissolved in polar solvents such as acetonitrile or 3-methoxy propionitrile. The counter-electrode is also fabricated by deposition of platinum thin film on FTO or ITO substrates and subsequently the cell is completed by attachment of both electrodes via a thermoplastic polymer. DSC cells are fabricated as large modules. **Figure 7** shows the system boundaries for an example of DSC LCA study.

Few LCA studies have been conducted on DSCs. Greijer et al. [60] evaluated LCA on a DSC system under 2190 kWh/m^2/year solar irradiation. The resulted GHG emissions have been ranged from 19 to 25 g $CO_{2\text{-eq}}$/kWh for DSC module efficiencies ranging between 7 and 12%.

Parisi et al. [61] realized lab-scale environmental analysis on DSC manufacturing. This study assumed efficiency of DSC module as 8% and the module's lifetime assumed 20 years. They demonstrated that the NER, EPBT and CO2 emission values as 12.64, 1.58 years and 22.38 g $CO_{2\text{-eq}}$/kWh, respectively. The authors also compared the LCA results of DSC with some typical thin film PV technologies and showed lower GHG emission and NER values for DSC compared with other organic and inorganic thin films. In another research, Parisi et al. [62] have been reported further LCA on DSCs from previous studies and compared the DSC with other thin film technologies (i.e., polymeric, a-Si, CdTe, and CIS). The environmental impact analysis reported in their work has been performed by using the ReCiPe2008 endpoint level approach associated with a hierarchist perspective. 17 category indicators were Fossil Depletion, Metal Depletion, Agricultural Land Occupation, Natural Land Transformation, Freshwater Ecotoxicity, Freshwater Eutrophication, Urban Land Occupation, Marine Ecotoxicity, Terrestrial Ecotoxicity, Terrestrial Acidification, Climate Change Ecosystems, Ionizing Radiation, Particulate Matter Formation, Photochemical Oxidant Formation, Ozone Depletion, Human Toxicity and Climate Change Human Health. 10 DSC configurations considered in their evaluation (**Table 2**) which in general the DSC with double PET structure show lower environmental impact (**Figure 8**). These results show the effect of PV substrate on environmental life cycle impact, in addition to precursor material.

Based on previous evaluations from many researchers, the range CED, EPBT and GHG emissions rate of DSCs are 277–365 MJ/m^2, 0.6–1.8 years and 9.8–120.0 g $CO_{2\text{-eq}}$/kWh, respectively. It is notably that also the GHG emissions of DSC strongly related to the operational lifetime of DSC modules. The best performance of DSC could be obtained with higher conversion efficiencies and longer lifetimes.

Recently, single-junction polymer solar cells (OPVs) can attract the PV community attention due to possibility of roll-to-roll printing fabrication. Lizin et al. [63] focused on the environmental impact of OPVs. They used a P3HT/PC_{60}BM active layer blend processed on semi-industrial pilot lines which is working under ambient conditions. Both standard and inverted device structures have been used in their work. The general steps in the LCA framework were standardized according to the ISO 14000 series which is shown in **Figure 9**.

The environmental impact was found to be strongly decreased through continuous manufacturing procedures. The current top performing cell regarding environmental performance has an energy payback time in the order of months and a cumulative energy demand of 37.58 MJp m^2 for cells having 2% efficiency.

In the past few years, Quantum dot sensitized solar cell (QDSC) was introduced based on the DSC structure as an alternative dye because of its high absorption coefficient, photostability and low cost [64]. Various QD materials such as CdS, CdTe, CdSe, ZnSe, PbS and InAs have been used as sensitizer to absorb visible light of solar irradiation. The typical structure of QDSC is similar to DSC, which

Configuration	Abbreviation
Glass–liquid electrolyte–platinum–glass	A
Glass–ionic liquid (type 3)–platinum–glass	B
Glass–ionic liquid (type 6)–platinum–glass	C
Glass–ionic liquid(type7)–platinum–glass	D
Steel–liquid electrolyte–platinum–glass	E
Steel–liquid electrolyte–platinum–PET	F
PET–liquid electrolyte–platinum–glass	G
PET–liquid electrolyte–platinum–PET	H
Glass–ionic liquid(type1)–cobalt sulfide–PET	I
Glass–liquid electrolyte–carbon	L

Table 2.
DSC configuration of LCA analysis obtained from ref [62].

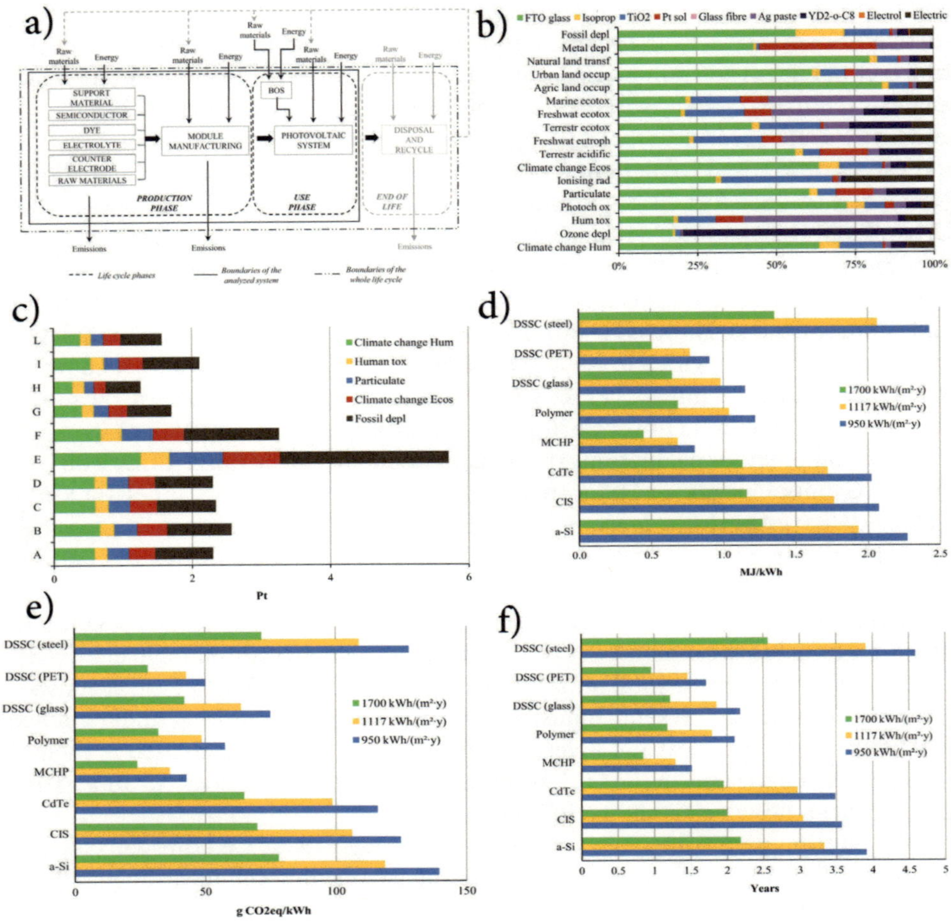

Figure 8.
(a) Designed boundaries of LCA evaluation of DSCs. (b) LCA characterization diagram of DSC module (containing YD2-o-C8 dye and CoII/CoIII redox couple). (c) Diagram of ReCiPe2008 single score analysis for DSC modules configurations. The calculated CED (d), GWP (e) and EPBT (f) indicators for kWh electricity production of different PV modules. The figures are obtained under copyright of Elsevier [62].

consists of counter and photoanode electrodes, but the only difference is that dye is replaced with QD materials in this structure [65]. Different techniques have been utilized to produce QD sensitizers coated on mesoporous TiO_2 either by in-situ or ex-situ fabrication [66]. There are two well known in-situ fabrication methods, (1) Chemical Bath Deposition (CBD) and (2) Successive Ionic Layer Adsorption and Reaction (SILAR). These methods have been generally used due to applicability in large-scale production [66]. Roll to-roll coating process has been also introduced for fabrication of large scale QDSC, before module encapsulation and framing.

Sengül and Theis evaluated LCA on a proposed type of nanostructure, QDPV module [67] using a roll-to-roll manufacturing process (**Figure 10**). The LCA parameters evaluated and compared with other PV and energy resources. Furthermore, emission of NO_x and SO_x gases and several heavy metals such as, mercury, nickel, arsenic, chromium, cadmium, and lead are evaluated for QDPVs and compared with other PV technologies. Their results indicate that while QDPV modules have shorter EPBT, lower Global Warming Potential (GWP), SO_x and NOx emissions than other types of PV modules, they have higher heavy metal emissions.

As a flagship of emerging thin film technologies, Perovskite Solar Cells (PSCs) obtain a remarkable increase of the power conversion efficiency (PCE) leading

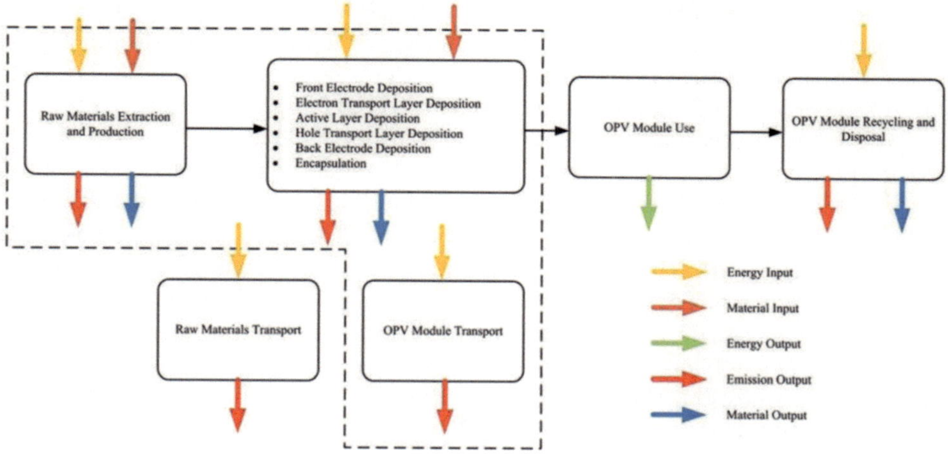

Figure 9.
OPV life cycle analysis scope definition the figure is obtained under copyright of Royal Society of Chemistry [63].

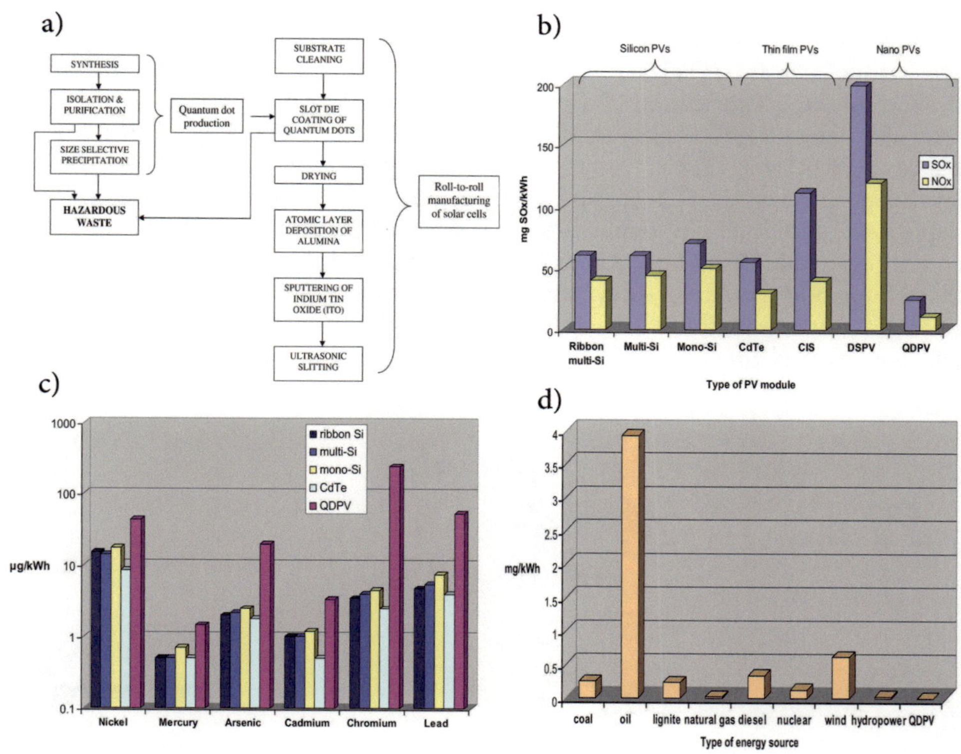

Figure 10.
(a) A typical quantum dot solar cell production process. (b) Comparison of emissions of sulfur oxides and nitrogen oxides for QDPV modules with different types of PV modules. (c) Comparison of heavy metal emissions from QDPV modules with silicon and CdTe PV modules. (d) Total amount of heavy metal emissions from QDPV modules compared to heavy metal emissions from other types of energy sources. The figures are obtained under copyright of Elsevier [67].

to a certified efficiency of 24.2% [4]. The excellence of PSCs marks a revolution among PVs by promising combination of low cost procedure (solution based) and high efficiency while in the typical commercial PV modules, manufacturing of the semiconductor absorber layer is an energy intensive process. High recorded values of the power conversion efficiency (PCE) is related to outstanding characteristics of hybrid perovskite such as, long carrier lifetimes, broad absorption and strong diffusion lengths, defect tolerance, direct band-gap and low recombination. These outstanding properties of hybrid perovskite led to various applications [68–71]. PSCs usually consist of a conductive substrate (FTO or ITO), an electron selective layer (ESL), a perovskite absorber, a hole transporting material (HTM) and a back contact electrode (gold, silver or carbon). Many inorganic structures have been utilized as ESL such as metal oxides, including TiO_2 (mesoscopic and planar structures) [18], ZnO [72] and SnO_2 [73], as well as polyoxometalates [74] which can provide low-temperature fabrication in addition to high efficiencies. Deposition of the hybrid perovskite is one of the main challenges for a high-reproducible, atmosphere-independent PSC fabrication technique due to air-sensitive nature of perovskite layer. In particular, Yaghoobi Nia et al. have developed a Crystal Engineering (CE) approach for kinetically controlling the perovskite phase nucleation and subsequently crystal growth under ambient conditions, through the formation of a lead based intermediate phase, reaching high efficiency $MAPbI_3$-based PSCs (more than 17%) and perovskite solar modules (~13%) [18] (**Figure 11**) which represent a promising approach for future up-scaling and commercial programs of PSCs.

The system boundary of a cradle-to-grave LCA study for a typical PSC system is presented in **Figure 12a** [75].

Raw materials extraction is classified at first stage while preparation of the precursors, such as FTO glass, TiO_2, PbI_2, Au, and PET materials, to be defined in the second stage. Fabrication of the mentioned perovskite module consists of, (1) laser ablation of the FTO glass substrates to define the cell arrays, (2) deposition of blocking TiO2 layer, (3) electron transport layer deposition (ETL), (4) perovskite layer deposition on the ETL (n-i-p structure), (5) hole transport layer (HTL) deposition, and finally (6) deposition of gold by thermal evaporation as back contact electrode [75]. The modules have been encapsulated by PET polymers. After assembling of the modules, they have been utilized to generate electricity from sun light. At the end, the waste modules have been landfilled in the disposal stage without any recycling program.

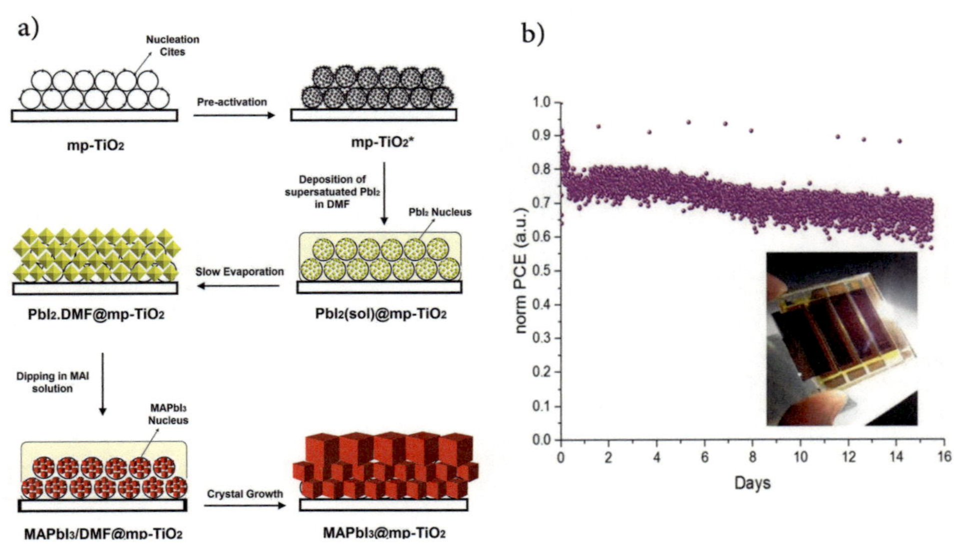

Figure 11.
(a) A schematic presentation of crystal engineering (CE) approach in order to fabrication perovskite solar modules under ambient condition. (b) Light-soaking stability test results of the presented PSC which is fabricated through CE. The figures are obtained under copyright of Royal Society of Chemistry [18].

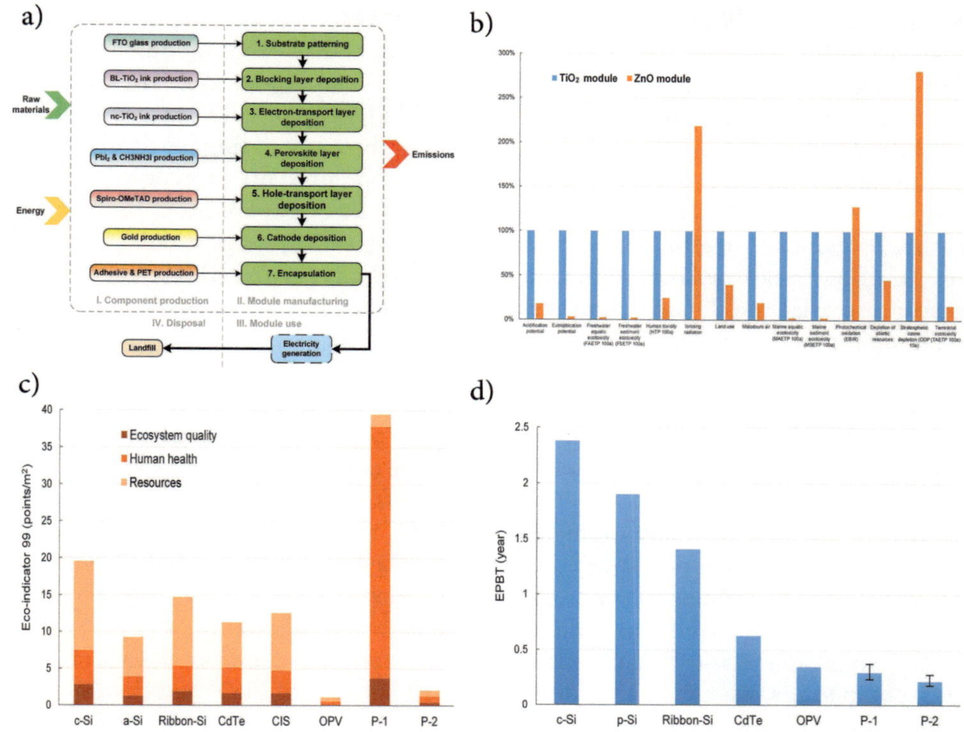

Figure 12.
(a) System boundary of manufacturing a perovskite solar module using TiO_2 as ETL. (b) Life cycle impact assessment comparison between 1 m^2 of the TiO_2 module and 1 m^2 of the ZnO module. (c) Eco-indicator 99 results for 1 m^2 of eight PV modules. P-1 represents the TiO_2 perovskite module; P-2 represents the ZnO perovskite module. (d) Energy payback time for seven PV modules. P-1 represents the TiO_2 perovskite module; P-2 represents the ZnO perovskite module. The figures are obtained under copyright of Royal Society of Chemistry [75].

Celik et al. [76] have been assessed cradle-to-gate the environmental impacts for three PSC modules consist of, vacuum, solution, and HTM-free structures, considering of 15% module efficiency, 15-year life span, and 65% active area under 1700 kWh/m^2/year solar irradiation. Owing to their results, EPBT and GHG emissions of the perovskite modules could be ranged between 1.05 and 1.54 years and 100 to 150 g $CO_{2\text{-eq}}$/kWh, respectively, which can be considered as lower values when compare with crystalline and commercial thin film technologies.

In particular, Gong et al. [75] have been evaluated LCA of two types of perovskite solar modules with two different ETLs (ZnO and TiO_2) by cradle to-grave approach to observe 16 LCIs and two sustainable indicators, namely, EPBT and GHG emission, under irradiation of 1960 kWh/m^2/year, assuming a lifetime of 2 years and PR of 0.80. The authors found that primary energy which is consumed by TiO_2 and ZnO modules are 446 MJ/m^2 and 392 MJ/m^2, respectively while GHG emission values calculated as 2.17 and 1.91 g $CO_{2\text{-eq}}$/m^2 respectively for TiO2 and ZnO modules.

The comparison of life cycle impact assessment results between the two modules is presented in (**Figure 12b**). It is evidenced that the ZnO module performs in a more environmental friendly manner except three impact categories: (1) photochemical oxidation, (2) ionizing radiation, and (3) stratospheric ozone depletion. Furthermore, in **Figure 12c** there is a comparison of Eco-indicator 99 results for eight types of PV modules. ZnO module has been achieved the second lowest level for three damage categories (ecosystem quality, resources and human health).

The Eco-indicator 99 values of the ZnO module are one order of magnitude lower than the values for c-Si, a-Si, ribbon-Si, CdTe, CIS, and the TiO_2 module. The EPBT comparison among seven PV modules is shown in **Figure 12d**. The ZnO perovskite module has been shown the lowest EPBT value of 0.22 years. The OPV module just shows a slightly longer EPBT values. The outstanding performance against silicon-based and CdTe modules is higher potential for a roll-to-roll process [77]. The EPBT of a perovskite module can be rapidly reduced in the near future by using more efficient processing technologies [18]. In particular, the LCA results of perovskite solar modules show the importance of the ETL layer in the case of material and deposition process which need to carefully considered in the scale-up research activities.

Summing up, the reports of LCA for perovskite solar cells show the EPBT varied from 0.2 to 5.4 years and the GHG emissions rate within 56.65–497.2 g $CO_{2\text{-eq}}$/kWh. Compared with silicon and thin film technologies, perovskite solar cells perform much lower energy consumption with competitive environmental benefits and EPBT which pave the way toward industrial manufacturing.

One of the important environmental concerns for perovskite solar cells is related to using of lead as an important element of active layer. Recently in an interesting activity, Yoo et al. [78] have been used Environmental Fate Modeling (EFM) model in order to evaluate of the lead compounds exposure in PSCs and their impact on the environment and humans ecosystem. Two main accidental situations of such compounds, have been considered as fire (formation of PbO) and flooding (formation of PbI_2). The results show water systems as the most unprotected to the toxicity of lead exposure. Accordingly, the assessments have been conducted into two categories, (1) upper limit suggestion for PSC production per unit area through the Predicted Environmental Concentrations (PEC) of PbI_2 and PbO, (2) lower limit suggestion of the demand safety management area through PEC evaluation in various environmental areas., PEC_{soil} values show ~10^3 less than the Predicted No Effect Concentration (PNEC) values for both compounds even using more than 100 MW power plant. The PEC_{soil} of PbI_2 is calculated as 1.4 ppm from 100 MW of PSC plants, while that of PbO is 3 ppm. An industrial power generation scale of 10 GW should exist in a 1 km^2 area to reach the concentrations near the PNEC values. Thus, the environmental risks on the soil concentrations could be classified as insignificant. However, the mentioned lead compounds can also penetrate into water resources. The risk factor of PbI_2 and PbO could be reached 1 when the PV plant scale reaches 21.8 and 6.39 MW km^{-2}, respectively. Therefore, the partitioning strategies of lead compounds per unit of PSC output generation plant, near water sources should be carefully considered for define the regulations by limiting the plant scale per unit area.

3. Conclusions

This chapter reviewed several previous environmental impact assessment studies on commercial and emerging thin film solar PV technologies including a-Si, CdTe, CIGS, DSC, PSC, OPV and QDSC. Three main indicators of LCA, such as CED, EPBT, and GWP, are considered and summarized. Across the analysis, emerging thin film PVs, especially perovskite solar cells, demonstrated the lowest energy requirement and shorter EPBT among other conventional PV technologies due to its high efficiency, low-cost production and eco-compatibility. Furthermore, the general life cycle assessment results show that the concern related to environmental impacts of heavy metal emission of some thin film PV technologies, particularly CdTe (emission of Cd) and PSCs (emission of Pb), is not noticeable when compare with life cycle of wafer based commercial PVs and typical fossil fuels. However, the environmental fate modeling results show that the effect of various environmental and human factors should be assessed and safety standards should be established using the most conservative range among various environmental evaluation results. Lastly, we conclude that, low-temperature solution-based produced perovskite solar modules are potentially the most environmentally sustainable PV if future development confirms a larger performance ratio and a longer lifetime.

Acknowledgements

Authors gratefully acknowledge the supports from Kimia Solar Co. and IRITALY Trading Company S.r.l.

Conflict of interest

The authors declare no conflict of interest.

Author details

Mahmoud Zendehdel[1*], Narges Yaghoobi Nia[2] and Mohammadreza Yaghoubinia[1,3]

1 K.S.R.I (Kimia Solar Research Institute), Kimia Solar Company, Kashan, Iran

2 CHOSE (Centre for Hybrid and Organic Solar Energy), University of Rome "Tor Vergata", Rome, Italy

3 Mobarakeh Steel Company, Mobarakeh, Esfahan, Iran

*Address all correspondence to: m.zendehdel@kimiasolar.com

References

[1] Demirbas MF. Electricity production using solar energy. Energy Sources Part A-recovery Utilization and Environmental Effects. 2007;**29**(6):563-569. DOI: 10.1080/009083190957685

[2] Jean J, Brown PR, Jaffe RL, Buonassisi T, Bulović V. Pathways for solar photovoltaics. Energy & Environmental Science. 2015;**8**(4): 1200-1219. Available from: http://xlink.rsc.org/?DOI=C4EE04073B

[3] Czochralski J. Ein neues verfahren zur messung der kristallisationsgeschwindigkeit der metalle scholar. Zeitschrift für Physikalische Chemie. 2017;**92U**(1):219-221. Available from: https://www.degruyter.com/view/j/zpch.1918.92.issue-1/zpch-1918-9212/zpch-1918-9212.xml

[4] National Renewable Energy Laboratory. Best Research-Cell Efficiency Chart|Photovoltaic Research|NREL [Internet]. Available from: https://www.nrel.gov/pv/cell-efficiency.html [Accessed: 04-March-2019]

[5] Yablonovitch E, Gmitter T, Harbison JP, Bhat R. Extreme selectivity in the lift-off of epitaxial GaAs films. Applied Physics Letters. 1987;**51**(26):2222-2224. DOI: 10.1063/1.98946

[6] Shah AV, Schade H, Vanecek M, Meier J, Vallat-Sauvain E, Wyrsch N, et al. Thin-film silicon solar cell technology. Progress in Photovoltaics: Research and Applications. 2004;**12**(23):113-142. DOI: 10.1002/pip.533

[7] Staebler DL, Wronski CR. Reversible conductivity changes in discharge-produced amorphous Si. Applied Physics Letters. 1977;**31**(4):292-294. DOI: 10.1063/1.89674

[8] Kessler F, Rudmann D. Technological aspects of flexible CIGS solar cells and modules. Solar Energy. 2004;**77**(6):685-695. Available from: https://www.sciencedirect.com/science/article/pii/S0038092X04000854

[9] Hetzer MJ, Strzhemechny YM, Gao M, Contreras MA, Zunger A, Brillson LJ. Direct observation of copper depletion and potential changes at copper indium gallium diselenide grain boundaries. Applied Physics Letters. 2005;**86**(16):162105. DOI: 10.1063/1.1906331

[10] Werner JH, Mattheis J, Rau U. Efficiency limitations of polycrystalline thin film solar cells: Case of Cu(In,Ga)Se2. Thin Solid Films. 2005;**480-481**:399-409. Available from: https://www.sciencedirect.com/science/article/pii/S0040609004016128

[11] Nishiwaki S, Siebentritt S, Walk P, Lux-Steiner MC. A stacked chalcopyrite thin-film tandem solar cell with 1.2 V open-circuit voltage. Progress in Photovoltaics: Research and Applications. 2003;**11**(4):243-248. DOI: 10.1002/pip.486

[12] Todorov TK, Tang J, Bag S, Gunawan O, Gokmen T, Zhu Y, et al. Beyond 11% efficiency: Characteristics of state-of-the-art $Cu_2ZnSn(S,Se)_4$ solar cells. Advanced Energy Materials. 2013;**3**(1):34-38. DOI: 10.1002/aenm.201200348

[13] Mendis BG, Shannon MD, Goodman MC, Major JD, Claridge R, Halliday DP, et al. Direct observation of Cu, Zn cation disorder in Cu_2ZnSnS_4 solar cell absorber material using aberration corrected scanning transmission electron microscopy. Progress in Photovoltaics: Research and Applications. 2014;**22**(1):24-34. DOI: 10.1002/pip.2279

[14] Kim H-S, Lee C-R, Im J-H, Lee K-B, Moehl T, Marchioro A, et al. Lead iodide Perovskite sensitized all-solid-state submicron thin film Mesoscopic solar cell with efficiency exceeding 9%. Scientific Reports. 2012;**2**(1):591. Available from: http://www.nature.com/articles/srep00591

[15] Eperon GE, Stranks SD, Menelaou C, Johnston MB, Herz LM, Snaith HJ. Formamidinium lead trihalide: A broadly tunable perovskite for efficient planar heterojunction solar cells. Energy & Environmental Science. 2014;7(3):982. Available from: http://xlink.rsc.org/?DOI=c3ee43822h

[16] Hao F, Stoumpos CC, Cao DH, Chang RPH, Kanatzidis MG. Lead-free solid-state organic–inorganic halide perovskite solar cells. Nature Photonics. 2014;**8**(6):489-494

[17] Noh JH, Im SH, Heo JH, Mandal TN, Seok S II. Chemical Management for Colorful, efficient, and stable inorganic–organic hybrid nanostructured solar cells. Nano Letters. 2013;**13**(4):1764-1769. DOI: 10.1021/nl400349b

[18] Yaghoobi Nia N, Zendehdel M, Cinà L, Matteocci F, Di Carlo A. A crystal engineering approach for scalable perovskite solar cells and module fabrication: A full out of glove box procedure. Journal of Materials Chemistry A. 2018;**6**(2):659-671. Available from: http://xlink.rsc.org/?DOI=C7TA08038G

[19] Liu M, Johnston MB, Snaith HJ. Efficient planar heterojunction perovskite solar cells by vapour deposition. Nature. 2013;**501**(7467):395-398. Available from: http://www.ncbi.nlm.nih.gov/pubmed/24025775

[20] Xing G, Mathews N, Sun S, Lim SS, Lam YM, Gratzel M, et al. Long-range balanced electron- and hole-transport lengths in organic-inorganic CH3NH3PbI3. Science. 2013;**342**(6156):344-347

[21] Ball JM, Lee MM, Hey A, Snaith HJ. Low-temperature processed meso-superstructured to thin-film perovskite solar cells. Energy & Environmental Science. 2013;**6**(6):1739. Available from: http://xlink.rsc.org/?DOI=c3ee40810h

[22] Peumans P, Yakimov A, Forrest SR. Small molecular weight organic thin-film photodetectors and solar cells. Journal of Applied Physics. 2003;**93**(7):3693-3723. DOI: 10.1063/1.1534621

[23] Li G, Zhu R, Yang Y. Polymer solar cells. Nature Photonics. 2012;**6**(3):153-161. Available from: http://www.nature.com/articles/nphoton.2012.11

[24] Krebs FC, Gevorgyan SA, Alstrup J. A roll-to-roll process to flexible polymer solar cells: Model studies, manufacture and operational stability studies. Journal of Materials Chemistry. 2009;**19**(30):5442. Available from: http://xlink.rsc.org/?DOI=b823001c

[25] Kim JY, Lee K, Coates NE, Moses D, Nguyen T-Q, Dante M, et al. Efficient tandem polymer solar cells fabricated by all-solution processing. Science. 2007;**317**(5835):222-225

[26] Günes S, Neugebauer H and, Sariciftci NS. Conjugated polymer-based organic solar cells. Chemical Reviews. 2007;**107**(4):1324-1338. DOI: 10.1021/cr050149z

[27] Kawano K, Pacios R, Poplavskyy D, Nelson J, Bradley DDC, Durrant JR. Degradation of organic solar cells due to air exposure. Solar Energy Materials & Solar Cells. 2006;**90**(20):3520-3530. Available from: https://www.sciencedirect.com/science/article/pii/S0927024806002960

[28] Lunt RR, Osedach TP, Brown PR, Rowehl JA, Bulović V. Practical roadmap

and limits to nanostructured photovoltaics. Advanced Materials. 2011;**23**(48):5712-5727. DOI: 10.1002/adma.201103404

[29] Hagfeldt A, Boschloo G, Sun L, Kloo L, Pettersson H. Dye-sensitized solar cells. Chemical Reviews. 2010;**110**(11):6595-6663. DOI: 10.1021/cr900356p

[30] Yella A, Lee H-W, Tsao HN, Yi C, Chandiran AK, Nazeeruddin MK, et al. Porphyrin-sensitized solar cells with cobalt (II/III)-based redox electrolyte exceed 12 percent efficiency. Science. 2011;**334**(6056):629-634. Available from: http://www.ncbi.nlm.nih.gov/pubmed/22053043

[31] Lan X, Masala S, Sargent EH. Charge-extraction strategies for colloidal quantum dot photovoltaics. Nature Materials. 2014;**13**(3):233-240. Available from: http://www.nature.com/articles/nmat3816

[32] Wang X, Koleilat GI, Tang J, Liu H, Kramer IJ, Debnath R, et al. Tandem colloidal quantum dot solar cells employing a graded recombination layer. Nature Photonics. 2011;**5**(8):480-484. Available from: http://www.nature.com/articles/nphoton.2011.123

[33] Chuang C-HM, Brown PR, Bulović V, Bawendi MG. Improved performance and stability in quantum dot solar cells through band alignment engineering. Nature Materials. 2014;**13**(8):796-801. Available from: http://www.nature.com/articles/nmat3984

[34] Zhitomirsky D, Voznyy O, Levina L, Hoogland S, Kemp KW, Ip AH, et al. Engineering colloidal quantum dot solids within and beyond the mobility-invariant regime. Nature Communications. 2014;**5**(1):3803. Available from: http://www.nature.com/articles/ncomms4803

[35] Ludin NA, Mustafa NI, Hanafiah MM, Ibrahim MA, Asri Mat Teridi M, Sepeai S, et al. Prospects of life cycle assessment of renewable energy from solar photovoltaic technologies: A review. Renewable and Sustainable Energy Reviews. 2018;**96**:11-28. Available from: https://www.sciencedirect.com/science/article/pii/S1364032118305574

[36] Godlee F. An international standard for disclosure of clinical trial information. BMJ. 2006;**332**(7550):1107-1108. Available from: http://www.ncbi.nlm.nih.gov/pubmed/16690647

[37] Yi S, Kurisu KH, Hanaki K. Life cycle impact assessment and interpretation of municipal solid waste management scenarios based on the midpoint and endpoint approaches. International Journal of Life Cycle Assessment. 2011;**16**(7):652-668. DOI: 10.1007/s11367-011-0297-3

[38] Rolf F, Garvin H, Marco R, Parikhit S, de Wild-Scholten M. Methodology Guidelines on Life Cycle Assessment of Photovoltaic Electricity. 3rd ed. Golden, CO (United States): National Renewable Energy Lab. (NREL); 2016. Available from: https://www.osti.gov/biblio/1351599

[39] Peng J, Lu L, Yang H. Review on life cycle assessment of energy payback and greenhouse gas emission of solar photovoltaic systems. Renewable and Sustainable Energy Reviews. 2013;**19**:255-274. Available from: https://www.sciencedirect.com/science/article/pii/S1364032112006478

[40] Pacca S, Sivaraman D, Keoleian GA. Parameters affecting the life cycle performance of PV technologies and systems. Energy Policy. 2007;**35**(6):3316-3326. Available from: https://www.sciencedirect.com/science/article/pii/S0301421506003715?via%3Dihub

[41] Minemoto T, Nagae S, Takakura H. Impact of spectral irradiance distribution and temperature on the outdoor performance of amorphous Si photovoltaic modules. Solar Energy Materials & Solar Cells. 2007;**91**(10):919-923. Available from: https://www.sciencedirect.com/science/article/pii/S0927024807000712

[42] Fthenakis VM. Life cycle impact analysis of cadmium in CdTe PV production. Renewable and Sustainable Energy Reviews. 2004;**8**(4):303-334. Available from: https://www.sciencedirect.com/science/article/pii/S1364032103001345?via%3Dihub

[43] Kato K, Hibino T, Komoto K, Ihara S, Yamamoto S, Fujihara H. A life-cycle analysis on thin-film CdS/CdTe PV modules. Solar Energy Materials & Solar Cells. 2001;**67**(1-4):279-287. Available from: https://www.sciencedirect.com/science/article/pii/S0927024800002932?via%3Dihub

[44] Ito M, Kato K, Komoto K, Kichimi T, Kurokawa K. A comparative study on cost and life-cycle analysis for 100 MW very large-scale PV (VLS-PV) systems in deserts using m-Si, a-Si, CdTe, and CIS modules. Progress in Photovoltaics: Research and Applications. 2008;**16**(1):17-30. DOI: 10.1002/pip.770

[45] Fthenakis VM, Kim HC. CdTe photovoltaics: Life cycle environmental profile and comparisons. Thin Solid Films. 2007;**515**(15):5961-5963. Available from: https://www.sciencedirect.com/science/article/pii/S0040609006016361?via%3Dihub

[46] Fthenakis VM, Kim HC, Alsema E. Emissions from photovoltaic life cycles. Environmental Science & Technology. 2008;**42**(6):2168-2174. DOI: 10.1021/es071763q

[47] Rocchetti L, Beolchini F. Recovery of valuable materials from end-of-life thin-film photovoltaic panels: Environmental impact assessment of different management options. Journal of Cleaner Production. 2015;**89**:59-64. Available from: https://www.sciencedirect.com/science/article/pii/S0959652614011809?via%3Dihub

[48] Bang Y-Y, Hong N-J, Sung Lee D, Lim S-R. Comparative assessment of solar photovoltaic panels based on metal-derived hazardous waste, resource depletion, and toxicity potentials. International Journal of Green Energy. 2018;**15**(10):550-557. DOI: 10.1080/15435075.2018.1505618

[49] Habibi MH, Karimi B, Zendehdel M, Habibi M. Fabrication, characterization of two nano-composite CuO-ZnO working electrodes for dye-sensitized solar cell. Spectrochimica Acta Part A. 2013;**116**:374-380

[50] Habibi MH, Habibi AH, Zendehdel M, Habibi M. Dye-sensitized solar cell characteristics of nanocomposite zinc ferrite working electrode: Effect of composite precursors and titania as a blocking layer on photovoltaic performance. Spectrochimica Acta Part A. 2013;**110**:226-232

[51] Hossein Habibi M, Askari E, Habibi M, Zendehdel M. Novel nanostructure zinc zirconate, zinc oxide or zirconium oxide pastes coated on fluorine doped tin oxide thin film as photoelectrochemical working electrodes for dye-sensitized solar cell. Spectrochimica Acta Part A. 2013;**104**:197-202

[52] Habibi MH, Mikhak M, Zendehdel M, Habibi M. Influence of nanostructured zinc titanate, zinc oxide or titanium dioxide thin film coated on fluorine doped tin oxide as working electrodes for dye-sensitized solar cell. International Journal of Electrochemical Science. 2012;7(8):6787

[53] Habibi MH, Karimi B, Zendehdel M, Habibi M. Preparation of nano-structure mixed copper-zinc oxide via co-precipitation rout for dye-sensitized solar cells: The influence of blocking layer and Co(II)/Co(III) complex redox shuttle. Journal of Industrial and Engineering Chemistry. 2014;**20**(4):1462-1467. Available from: https://www.sciencedirect.com/science/article/pii/S1226086X13003432

[54] Habibi MH, Mardani M, Habibi M, Zendehdel M. Enhanced photovoltage (Voc) of nano-structured zinc tin oxide (ZTO) working electrode prepared by a green hydrothermal route for dye-sensitized solar cell (DSSC). Journal of Materials Science: Materials in Electronics. 2017;**28**(4):3789-3795. DOI: 10.1007/s10854-016-5989-9

[55] Karimian D, Yadollahi B, Zendehdel M, Mirkhani V. Efficient dye-sensitized solar cell with a pure thin film of a hybrid polyoxometalate covalently attached organic dye as a working electrode in a cobalt redox mediator system. RSC Advances. 2015;**5**(94):76875-76882. Available from: http://xlink.rsc.org/?DOI=C5RA09104G

[56] Gong J, Liang J, Sumathy K. Review on dye-sensitized solar cells (DSSCs): Fundamental concepts and novel materials. Renewable and Sustainable Energy Reviews. 2012;**16**(8):5848-5860. Available from: https://www.sciencedirect.com/science/article/pii/S136403211200319X?via%3Dihub

[57] Su'ait MS, Rahman MYA, Ahmad A. Review on polymer electrolyte in dye-sensitized solar cells (DSSCs). Solar Energy. 2015;**115**:452-470. Available from: https://www.sciencedirect.com/science/article/pii/S0038092X15001139?via%3Dihub

[58] Yaghoobi Nia N, Farahani P, Sabzyan H, Zendehdel M, Oftadeh M. A combined computational and experimental study of the $[Co(bpy)_3]^{2+/3+}$ complexes as one-electron outer-sphere redox couples in dye-sensitized solar cell electrolyte media. Physical Chemistry Chemical Physics. 2014;**16**(23):11481-11491

[59] Nasr-Esfahani M, Zendehdel M, Yaghoobi Nia N, Jafari B, Khosravi BM. Fabrication and characterization of a new dye sensitized solar cell with a new Schiff base cobalt complex as a redox mediator. RSC Advances. 2014;**4**(31):1596

[60] Greijer H, Karlson L, Lindquist S-E, Hagfeldt A. Environmental aspects of electricity generation from a nanocrystalline dye sensitized solar cell system. Renewable Energy. 2001;**23**(1):27-39. Available from: https://www.sciencedirect.com/science/article/pii/S0960148100001117?via%3Dihub

[61] Parisi M, Sinicropi A, Basosi R. Life cycle assessment of gratzel-type cell production for non conventional photovoltaics from novel organic dyes. International Journal of Heat and Technology. 2011;**29**:161-169

[62] Parisi ML, Maranghi S, Basosi R. The evolution of the dye sensitized solar cells from Grätzel prototype to up-scaled solar applications: A life cycle assessment approach. Renewable and Sustainable Energy Reviews. 2014;**39**:124-138. Available from: https://www.sciencedirect.com/science/article/pii/S1364032114005310?via%3Dihub

[63] Lizin S, Van Passel S, De Schepper E, Maes W, Lutsen L, Manca J, et al. Life cycle analyses of organic photovoltaics: A review. Energy & Environmental Science. 2013;**6**(11):3136. Available from: http://xlink.rsc.org/?DOI=c3ee42653j

[64] Shen Y-J, Lee Y-L. Assembly of CdS quantum dots onto mesoscopic TiO_2 films for quantum dot-sensitized solar cell applications. Nanotechnology.

2008;**19**(4):045602. Available from: http://stacks.iop.org/0957-4484/19/i=4/a=045602?key=crossref.19ab08e1206f98636af4427396a516fd

[65] Sudhagar P, Jung JH, Park S, Sathyamoorthy R, Ahn H, Kang YS. Self-assembled CdS quantum dots-sensitized TiO2 nanospheroidal solar cells: Structural and charge transport analysis. Electrochimica Acta. 2009;**55**(1):113-117. Available from: https://www.sciencedirect.com/science/article/pii/S0013468609010561?via%3Dihub

[66] Jun HK, Careem MA, Arof AK. Quantum dot-sensitized solar cells—Perspective and recent developments: A review of Cd chalcogenide quantum dots as sensitizers. Renewable and Sustainable Energy Reviews. 2013;**22**:148-167. Available from: https://www.sciencedirect.com/science/article/pii/S1364032113000610?via%3Dihub

[67] Şengül H, Theis TL. An environmental impact assessment of quantum dot photovoltaics (QDPV) from raw material acquisition through use. Journal of Cleaner Production. 2011;**19**(1):21-31. Available from: https://www.sciencedirect.com/science/article/pii/S0959652610003252

[68] Salamandra L, Yaghoobi Nia N, Di Natali M, Fazolo C, Maiello S, La Notte L, et al. Perovskite photo-detectors (PVSK-PDs) for visible light communication. Organic Electronics. 2019;**69**:220-226. Available from: https://www.sciencedirect.com/science/article/pii/S1566119919301090?dgcid=rss_sd_all

[69] Taheri B, Yaghoobi Nia N, Agresti A, Pescetelli S, Ciceroni C, Del Rio Castillo AE, et al. Graphene-engineered automated sprayed mesoscopic structure for perovskite device scaling-up. 2D Materials. 2018;**5**(4):045034. Available from: http://stacks.iop.org/2053-1583/5/i=4/a=045034?key=crossref.2b5fac3fb2b987e29999f38b0cdf07d2

[70] Nia NY, Matteocci F, Cina L, Di Carlo A. High-efficiency perovskite solar cell based on poly(3-hexylthiophene): Influence of molecular weight and mesoscopic scaffold layer. ChemSusChem. 2017;**10**(19):3854-3860. DOI: 10.1002/cssc.201700635

[71] Yaghubinia M, Ebnali M, Mahmoud Zendehdel MY. Improvement of perovskite solar cells photovoltaic performance by localized surface plasmon effect of silver-alumina core-shell nanoparticles. 4th International Conference on Photonics, Optics and Laser Technology (PHOTOPTICS), In: Photoptics. IEEE; 2016. pp. 1-3. Available from: https://ieeexplore.ieee.org/abstract/document/7951303

[72] Song J, Zheng E, Liu L, Wang X-F, Chen G, Tian W, et al. Magnesium-doped zinc oxide as electron selective contact layers for efficient perovskite solar cells. ChemSusChem. 2016;**9**(18):2640-2647. DOI: 10.1002/cssc.201600860

[73] Liu Q, Qin M-C, Ke W-J, Zheng X-L, Chen Z, Qin P-L, et al. Enhanced stability of perovskite solar cells with low-temperature hydrothermally grown SnO_2 electron transport layers. Advanced Functional Materials. 2016;**26**(33):6069-6075. DOI: 10.1002/adfm.201600910

[74] Sardashti MK, Zendehdel M, Nia NY, Karimian D, Sheikhi M. High efficiency $MAPbI_3$ perovskite solar cell using a pure thin film of polyoxometalate as scaffold layer. ChemSusChem. 2017;**10**(19):3773-3779

[75] Gong J, Darling SB, You F. Perovskite photovoltaics: Life-cycle assessment of energy and environmental impacts. Energy & Environmental Science. 2015;**8**(7):1953-1968.

Available from: http://xlink.rsc.org/?DOI=C5EE00615E

[76] Celik I, Song Z, Cimaroli AJ, Yan Y, Heben MJ, Apul D. Life cycle assessment (LCA) of perovskite PV cells projected from lab to fab. Solar Energy Materials & Solar Cells. 2016;**156**:157-169. Available from: https://www.sciencedirect.com/science/article/pii/S0927024816300605?via%3Dihub

[77] Krebs FC, Espinosa N, Hösel M, Søndergaard RR, Jørgensen M. 25th Anniversary article: Rise to power—OPV-based solar parks. Advanced Materials. 2014;**26**(1):29-39. DOI: 10.1002/adma.201302031

[78] Yoo YG, Park J, Umh HN, Lee SY, Bae S, Kim YH, et al. Evaluating the environmental impact of the lead species in perovskite solar cells via environmental-fate modeling. Journal of Industrial and Engineering Chemistry. 2019;**70**:453-461. Available from: https://www.sciencedirect.com/science/article/pii/S1226086X1831339X

Section 3

Ecological Concerns

Chapter 7

Toxic Materials Used in Thin Film Photovoltaics and Their Impacts on Environment

Hervé Joël Tchognia Nkuissi, Fransisco Kouadio Konan, Bouchaib Hartiti and Jean-Marie Ndjaka

Abstract

Photovoltaic industry has proved to be a growing and advantageous source of energy as it can be renewable, sustainable, reliable and clean. Significant improvements have been made in materials used and the production processes to reduce the costs, and to avoid possible issues induced by some hazardous materials. However, some health and environment challenges last, which must be overcome to make this technology a source of truly clean energy. This chapter provides an overview on the major environmental impacts of thin film technology associated with the use of toxic materials and the chemicals in the manufacturing processes. A summary of Environmental, Health and Safety issues associated with some thin film technologies like copper indium gallium diselenide (CIS/CIGS), cadmium telluride (CdTe) and amorphous silicon (a-Si) is done, in order to investigate potential infections induced by the environmental release of trace elements, usually coming from chemical vapor inhalation and eventually accidental spills during the manufacturing processes, on the health of humans and animals. Potential solutions will be provided to prevent some environmental issues.

Keywords: hazardous materials, toxic chemicals, manufacturing processes, environmental impacts, thin film technology, recycling, waste minimization

1. Introduction

The sun undoubtedly is known to be an incredible and inexhaustible source of energy, allowing the generation of electricity and showing distinct environmental advantages over conventional source. Once took out from the manufactory, photovoltaic (PV) systems do not produce any toxic gas emissions, any noise or greenhouse gases. However, as with any industrial product, there are health and environmental impacts associated with the manufacture of solar cells and solar panels. The PV industry uses harmful and flammable substances, although in small amounts, which can involve environmental and occupational risks. The main environmental impacts of solar panels are associated with the use of land, water, natural resources, hazardous materials, life-cycle global warming emissions etc.

The solar cell manufacturing process involves a number of harmful chemicals. These substances, similar to those used in the general semiconductor industry, include sulfuric acid, hydrogen fluoride, hydrochloric acid, nitric acid, 1,1,1-trichloroethane, and acetone. The amount and type of chemicals used depends on the type of cell and the technology used [1]. Thin film PV (TFPV) technology contains a higher number of toxic materials than those used in traditional silicon PV technology, including indium, gallium, arsenic, selenium, cadmium, telluride [2]. These materials must be handled and disposed of properly, to avoid with time serious environmental and human health problems.

The current idea of the industry is to ensure that these highly valuable and often rare materials are recycled, to foresee the pollution hazards. This chapter deals with the possibility of chemicals used in PV cell manufacturing process to be released to air, water surface and the environment. An overview on the TFPV industry will be done to understand how TFPV cells and modules are designed and fabricated. Some hazardous materials and chemicals used in the manufacture of TFPV technology and their relative toxicity to human health and environment will be produced. Finally, some solutions to anticipate long term harmful impacts of these products will be proposed.

2. Overview of the thin film photovoltaic industry

2.1 Development of thin film solar cells

The ultimate goal in the manufacturing of a PV module is widely determined by the cost per unit power output. The development of new PV technologies based on thin film materials has been led by the need for cheaper and more efficient semiconductor materials. Thin film solar cells (TFSCs) have the potential for rapid growth and low cost production. They have several advantages in manufacturing processes compared to conventional silicon solar cells such as [3]:

- Cheaper to produce
- Lower consumption of materials
- Fewer processing steps
- Availability of materials
- Simplified materials handling
- Can be deposited on many different substrates
- A variety of deposition technique
- Process lends itself to automation
- Integrated, monolithic circuit design instead of assembly of individual solar cells into final products

TFSCs are typically made up of thin layers of semiconductor materials, for instance cadmium or zinc sulfide, glass, and a contact material. The materials used in the design of TFSCs include polycrystalline silicon, amorphous silicon (a-Si), and semiconductors compounds. Semiconductors compounds include cadmium sulfide (CdS), cadmium telluride (CdTe), copper sulfide (CuS_2), copper indium diselenide (CIS), copper indium gallium diselenide (CIGS), copper gallium diselenide (CGS), germanium (Ge), and gallium arsenide (GaAs) [4]. Other semiconductors compounds like copper zinc tin sulfide (CZTS), copper zinc tin selenide (CZTSe)

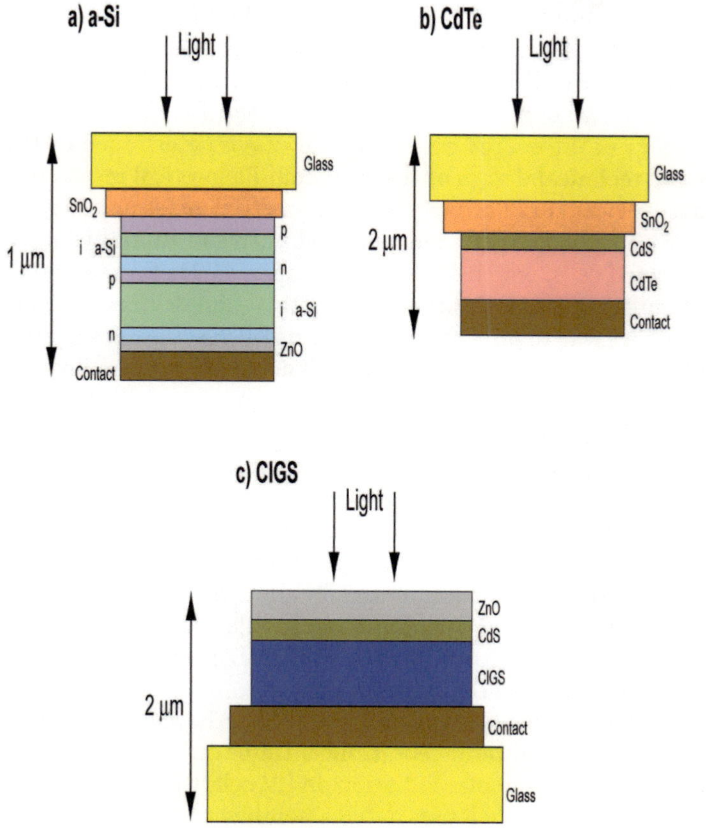

Figure 1.
Examples of thin-film solar cells structures: (a) amorphous silicon, (b) cadmium telluride, and (c) copper indium gallium diselenide [5].

and copper iron tin sulfide (CFTS) have proved over recent years their potential to convert the solar radiation into electricity and are being developing in laboratories. In these solar cells, the n material can be made of CdS or ZnS, while the p material can be made of $CuInSe_2$ (CIS) or Cu_2ZnSnS_4 (CZTS). Gallium arsenide (GaAs) solar cells can use aluminum, indium, or phosphorous as p or n-type materials. In **Figure 1**, are shown typical traditional structures of a-Si, CdTe and CIGS thin film solar cells.

To achieve better conversion efficiency, the active layer of the cell should have its band gap energy within the optimum range of 1.1–1.8 eV [5, 6]. Amorphous silicon is just at the rear end of the optimum band gap range at 1.8 eV [6]. Ge (0.66 eV) and CdS (2.45 eV) for example, have band gap energies outside of the optimum range, which means that those materials show limited conversion efficiencies. Ge is usually used to improve conversion efficiency of amorphous silicon cells [7], whereas, CdS is used to improve conversion efficiency in CIS and CdTe cells [8]. In general, on the basis of the band gap, materials used in TFSCs such as GaAs and CdTe have higher theoretical conversion efficiency than crystalline silicon, as they show band gap energies close to the optimum value of approximately 1.5 eV [8].

Thin film materials have higher light absorption capabilities than crystalline silicon as they have a direct absorption profile. Therefore, they can be shaped thinner than conventional silicon which must cut on wafers. For example, 1 µm of a GaAs direct semiconductor is sufficient to absorb the same quantity of the photons light than 100 µm of an indirect silicon semiconductor. CIS or CIGS are direct band gap polycrystalline materials with high absorption coefficients in the order of 10^5 cm^{-1}, allowing the active layer only to be about 2 µm [9]. Some CIGS-based solar cells usually introduce a thin film of CdS on the top of the CIGS layer as a buffer layer, in order to ensure the electrical transition between the CIGS layer and the window layer (the front of the cell). Zinc, molybdenum, tin, and aluminum are used in these cells as front and back contacts or components of the layers.

The efficiency of the energy conversion process is determined by the materials parameters and technical design of the solar cell. Theoretical research on TFSCs indicates that, devices could achieve conversion efficiencies up to 25% under lab conditions [10]. In addition, efficiencies of cells can substantially increase by stacking interconnected cells, which could achieve up to 41.9% for tandem (two cells) and 50% for multi cells [11]. Concentrator systems and devices used to track the sun can be another way to improve efficiency of PV systems. For example, the use of terrestrial concentrator GaAs/Ge solar cells achieved conversion efficiency up to 36.9% [12]. The first TFSCs had conversion efficiencies of 8–12% [4]. **Table 1** presents the record lab efficiencies for cells and modules of different technology. Note that these are just record lab efficiencies, not commercially guaranteed efficiencies.

Crystalline silicon-based technologies continue to dominate the world market share with about 95% of the total production in 2017 [26]. The share of multi crystalline technology is now about 62% of total production [26]. In 2017, the market share of all thin film technologies counted for about 5% [26]. Amorphous silicon holds 4% of this part followed by cadmium telluride with 1%. The others, especially CIGS and thin film silicon, although already available at the commercial stage still represent a negligible part of the market.

Thin film materials such GaAs, $GaInP_2$, and CIGS have been investigated for the development of concentrator cells. Concentrator cells have been designed to increase the intensity of the solar radiation on PV cells through the use of optical lenses. They consist of optical lenses, a cell assembly, a housing element, a secondary concentrator to reflect off-center light rays onto the cell, a mechanism to dissipate excess heat produced by concentrated sunlight, and various adhesives and contacts [27]. The main advantages of concentrator cells are: they reduce the number or size of solar cells used, enhance the power output, and enhance the solar cell efficiency under concentrated sunlight [8]. A conversion efficiency of 32% has been reported for concentrator cells [28]. This other way to increase the cells efficiency nevertheless presents some drawbacks: they involve expensive tracking systems and

Technology	Cell efficiency (%)	Module efficiency (%)	Description cell/module
Crystalline silicon	26.7 ± 0.5	24.4 ± 0.5	Kaneka [14]/Kaneka (10^8 cells) [14]
Multi crystalline silicon	22.3 ± 0.4	19.9 ± 0.4	FhG-ISE [15]/Trina solar (120 cells) [16]
CIGS	21.7 ± 0.5	19.2 ± 0.5	Solar Frontier [17]/Solar Frontier (70 cells) [18]
CdTe	21.0 ± 0.4	18.6 ± 0.5	First Solar [19]/First Solar [20]
Thin film silicon	10.5 ± 0.3	–	CSG solar [21]/
Amorphous silicon	10.2 ± 0.3	9.1	AIST [22]
GaAs	28.8 ± 0.9	25.1 ± 0.8	Alta devices [23]/Alta devices [24]
CZTS	10.0 ± 0.2	–	UNSW [25]

Table 1.
Confirmed terrestrial record cell and module efficiencies measured under the global AM 1.5 spectrum (1000 W/m²) at a cell temperature of 25°C (IEC 60904-3: 2008, ASTM G-173-03 global) [13].

more precise controls than the traditional flat plate systems, they generate higher operating temperatures, which can decrease the long-term stability and lifetime of the PV cells. Concentrator cells were first designed for space applications, but modules for terrestrial applications are already commercially available [8].

2.2 The manufacturing processes

It is known that the complexity of solar cells and modules manufacturing strongly raises their costs. Conventional silicon is handled in different many ways, complicating therefore fabrication processes. First, silicon raw material is melted at very high temperatures and grown into a silicon ingot. Then, the ingot is molded and sawn into individual wafers for cell processing [3]. After testing, individual cells are connected together in a suitable electrical configuration. Finally, the connection circuit is hermetically packaged in a weatherproof flat container, typically with an aluminum frame. **Figure 2** shows the flowchart describing the complete process to manufacture a conventional crystalline silicon-based module. The process requires more than 20 separate steps before a module is complete.

In contrast to crystalline silicon, thin film manufacturing steps are very simple. For example, the connection of the circuit from individual cells is removed. Instead of processing and handling ingots, wafers and cells, the final circuit is directly fabricated on a single large substrate, usually glass [3]. Thin film circuits require

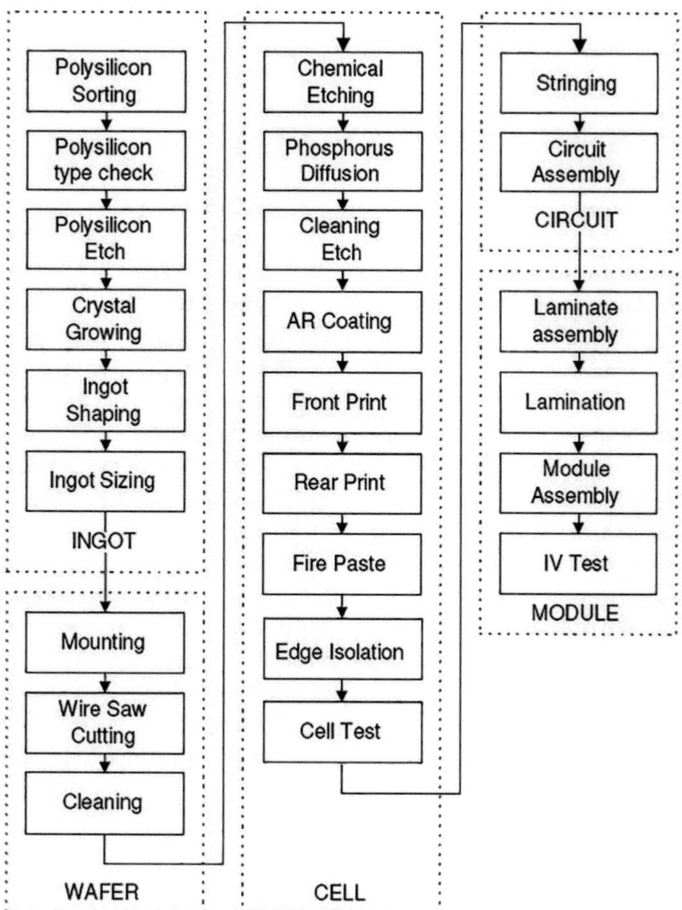

Figure 2.
Different process steps for fabrication of crystalline silicon modules [3].

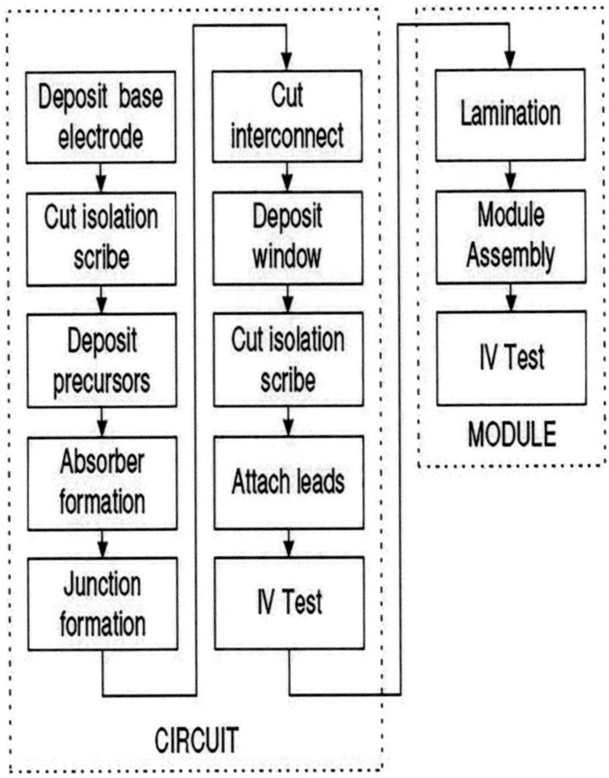

Figure 3.
Different process steps for fabrication of thin film modules [3].

the deposition of three main layers namely a back contact or a base electrode layer usually deposited on glass, a semiconductor layer and a transparent window layer also acting like a conductor front contact as it is typically seen in **Figure 1(c)**. The semiconductor layer is divided into the absorber and buffer layers. The stack of the absorber layer in one side and the buffer and window layers in another side creates an efficient n-p PV heterojunction. In TFSCs, the crucial phenomena of charge carriers generation and separation occur within the absorber; this layer therefore plays an important role in defining the electrical output parameters of the solar cell and usually confers its name to the technology. We have for instance CdTe for cadmium telluride based PV technology, CIGS for a range of chalcopyrite based PV technology and CZTS for a range of kesterite-based PV technology [3, 29].

An important advantage of thin film PV module manufacturing is found in the monolithic series interconnection of individual cells. Thin film cells are interconnected through simple patterning steps integrated into the processing line. The patterning steps achieve the integrated series interconnection from cell to cell on the circuit as shown in **Figure 3**. Three scribes between deposition steps complete the cell definition, separation and interconnection. A transparent conductive oxide (TCO) can also be integrated to the system for photon absorption optimization [29].

3. Chemicals and materials used in the fabrication of thin film cells and modules

To produce thin film PV devices, a variety of chemicals and materials is used. The types and quantities of chemicals used will depend on the type of the

CIS	CIGS	CZTS	GaAs	CdTe	Cu_2S	a-Si
Cadmium	Cadmium	Cadmium	Arsenic	Cadmium chloride	Ammonium chloride	Acetone
Copper	Copper	Copper	Arsine	Cadmium	Ammonium fluoroborate	Aluminum
Hydride gas	Gallium	Hydrogen selenide	Gallium	Molybdenum	Cadmium sulfide	Chloro-silanes
Hydrogen sulfide	Indium	Hydrogen sulfide	Hydrochloric acid	Nickel	Chromate coating	Diborane
Hydrogen selenide	Molybdenum	Molybdenum	Methane	Sulfur	Copper	Hydrochloric acid
Indium	Selenium	Selenium	Phosphine	Tellurium	Cuprous chloride	Hydrofluoric acid
Molybdenum	Zinc	Thiourea	Trichloroethylene	Thiourea	Gold	Hydrogen
Selenium		Tin	Triethyl gallium	Tin	Hydrochloric acid	Isopropanol
Zinc		Zinc	Trimethyl gallium		Hydrogen sulfide	Nitrogen
					Methanol	Phosphine
					Nickel	Phosphoric acid
					Nitrogen	Silane
					Polyvinyl butyral	Silicon tetrafluoride
					Silicon monoxide	Silicon
					Sodium chloride	Sodium hydroxide
					Tantalum pentoxide	Tin
					Zinc	
					Zinc fluoroborate	

Table 2.
Chemicals and materials involved in the manufacturing process of different thin film PV technology [8].

technology and the type of cell being produced. One can found also some variability in the use of chemicals for producing the same type of PV solar cells by different PV manufacturers. This means that each manufacturer has its own recipe to produce a type of solar cell. **Table 2** gives a general list, but non-exhaustive of chemicals and materials used in the manufacturing of some TFSCs and modules.

In TFPV technology, only few amounts of semiconductor materials are necessary to produce thin or ultra-thin layers of a solar cell. The amounts of chemicals and materials used in the manufacturing of TFPV devices vary depending on the type of cell being produced. For example, the quantity of cadmium in a CIS PV module is evaluated at 0.04 g/m^2 and in a CdTe PV module at 5 g/m^2 [30]. Research allowed reducing significantly the amount of cadmium in PV devices by using light-trapping methods [8]. These methods have led to reduce the thickness of CdTe layers from 2 to o.5 μm, corresponding to 5.5 and 0.55 g/m^2 amount of cadmium, respectively [8]. A diversified number of acids and corrosive liquids are used sensibly in large quantities during the manufacturing processes. These chemicals, similar to those used in the general semiconductor industry, and including sulfuric acid, hydrochloric acid, hydrogen fluoride and nitric acid are primarily used for cleaning wafers in the case of crystalline silicon or for removing impurities from raw

Material	Source	DOT hazard classification	Critical effects
Arsenic	GaAs	Poison	Cancer, lung
Arsine	GaAs (CVD)	Highly toxic gas	Blood, kidney
Cadmium	CdTe, CdS, $CdCl_2$	Poison	Cancer, kidney, bone
Diborane	a-Si dopant	Flammable gas	Pulmonary
Diethyl silane	a-Si deposition	Flammable liquid	
Diethyl zinc		Pyrophoric liquid	
Dimethyl zinc		Spontaneously combustible	
Hydrochloric acid	a-Si, GaAs, Cu_2S/CdS	Corrosive material	
Hydrofluoric acid	a-Si	Corrosive material	
Hydrogen	a-Si	Flammable gas	Fire hazard
Hydrogen selenide	CIS	Highly toxic gas	Irritant
Hydrogen sulfide	CIS, Cu_2S/CdS	Flammable gas	Irritant, Fire hazard
Indium	CIS, CIGS	Not regulated	Pulmonary, bone
Methane	GaAs	Flammable gas	Fire hazard
Molybdenum hexafluoride		Toxic and corrosive gas	
Oxygen	x-Si	Gaseous oxidizer	
Phosphine	a-Si dopant	Highly toxic and pyrophoric gas	Irritant, fire hazard
Phosphorus oxychloride	x-Si	Corrosive material	Irritant, kidney
Selenium	CIS, CZTS	Poison	Irritant
Silane	a-Si deposition	Pyrophoric gas	Irritant, fire, explosion hazard
Silicon tetrafluoride	a-Si deposition	Toxic and corrosive gas	

Material	Source	DOT hazard classification	Critical effects
Tellurium	CdTe	Not regulated	Cyanosis, liver
Tertiarybutyl arsine		Pyrophoric and highly toxic liquid	
Tertiarybutyl phosphine		Pyrophoric liquid	
Trimethyl aluminum		Pyrophoric liquid	
Trimethyl gallium	GaAs	Pyrophoric liquid	
Tungsten hexafluoride		Toxic and corrosive gas	

Table 3.
Hazard classification of chemicals typically used in PV module manufacturing [31, 32].

semiconductor materials. Solvents like acetone, ethanol and 1,1,1-trichloroethane are also used for cleaning in different steps of the fabrication processes.

Many hazardous materials as well as explosive and toxic gases are involved in the manufacturing processes of thin film PV cells and modules. **Table 3** presents a general list of some materials and chemicals and their description, classified as hazardous by the Department of Transportation (DOT) in USA and used in the whole PV industry. But the amounts and recipes vary from one manufacturer to another. Moreover, it is possible that some of these chemicals may no longer be used for PV devices production as the fabrication processes are constantly changing and evolving [8].

The wastes generated by the semiconductors materials used in TFPV industry are in general non-negligible. Acids and solvents each represented about one-third of the total wastes by weight (about 7000 tons) [33]. About 35% of the semiconductor wastes were evacuated as diluted acid solutions to sewage treatment plants and 37% were sent to offsite treatment facilities. About 27% of the total wastes were released to the atmosphere. Only 0.8% of the total wastes were discharged directly to the surface water and 0.015% to the landfills [33].

4. Potential health and environmental hazards

The manufacturing of PV devices includes some chemicals which can be toxic or harmful to the humankind. The potential for health concerns is not only depend on the material harmful characteristics, but also on certain conditions which must be taken into account. For example, in addition to harmful characteristics of the chemicals, their concentration must be high enough to constitute a real problem in a given environment: a human or an animal must be in the surroundings of where the device or compound is used; there must be a total exposition process from the compound to the environment. Most often, the primary persons exposed to the PV manufacturing residues are the plant workers. The easiest exposure route for workers is inhalation of vapors or dusts and also via direct contact if spills occur [8]. Another route for workers to be infected by chemicals resulting from manufacturing processes could be accidental ingestion. The ones outside a manufacturing environment could be infected by chemicals via inhalation from stack emissions, elusive air emissions or from accidental release after fire or explosion [8]. But the exposure of nearby residents or other workers would be less than the plan workers because the chemicals would be dispersed in the ambient air after their emission. There are possibilities for lands containing spent PV modules to pollute the environment. For example, at the surroundings of spent PV modules, groundwater seepage could reach a drinking water source or river; but in both cases, there would be dilution (not totally) of the waste before the water was used. In short, any vapor emissions or groundwater seepage would be diluted by the ambient air or by the water before reaching nearby residents.

It is well known that the fabrication of PV cells and modules needs the use of more or less large quantities of solvents and acids for synthesis and cleaning, gases for depositing ultra-thin film of semiconductors materials and metals according to the type of PV cell or module being fabricated. Most of these chemicals are highly toxic and harmful for humans and environment. Here are discussed some health and environment issues caused by chemicals hazards related to materials' toxicity, flammability, explosiveness, and carcinogen nature. Below is a summary of potential health and environmental issues concerning the manufacture and the use of some thin film technologies such as CdTe, a-Si and CI(G)S.

4.1 Cadmium telluride (CdTe)

The manufacturing of CdTe solar cells can cause occupational health risks associated with the toxicity of the main constitutive materials such as CdTe, CdS, and cadmium chloride ($CdCl_2$). Since cadmium compounds are usually used in powder and in liquid form, the primary route of exposure in manufactory settings is inhalation of cadmium-containing vapors or dust or ingestion of spills if this occurs. Processes in which cadmium compounds are used or produced in the form of fine fumes or particles present more risks to health, because they promote the absorption of these fine particles by the lung and thus can cause lung cancer. A long-term exposure can also have harmful effects on bone and kidney [32]. In addition, the inhalation of cadmium-containing vapors or dusts can result in metal vapor fever, pneumonitis, pulmonary edema, and finally death [32]. Since cadmium is produced primarily as a by-product of zinc mining, the levels of Cd production is fixed by the levels of zinc production. Because Zn is produced in large amounts, considerable amounts of Cd are also generated as by-product, without taking into account the amount used or required in PV technology. If the amount of Cd generated as a by-product of zinc is not totally absorbed by the whole market, it is discharged to the environment as hazardous waste. Thus, encapsulating Cd in CdTe for PV modules fabrication could be a trusty way to preserve the environment of hazards that can cause free elemental Cd. CdTe is more stable and insoluble to water; as such, it may be less toxic or harmful than free elemental Cd.

CdTe and CdS thin films are solid and are packaged into thick layers of glass or a waterproof container. At ambient conditions, the vapor pressure of CdTe is zero. Therefore, it is impossible for any vapors or dust to be released when using CdTe PV modules. The only or the more plausible way for cadmium to be released and absorbed by residents is via consumed modules in residential fires. Even in this way, flame temperatures in residential fires typically 800–1000°C, are not sufficient to vaporize CdTe [34]. The melting point of CdTe is 1041°C, and evaporation starts at 1050°C. The melting point of CdS is 1750°C [34]. Previous studies showed that CdTe releases are not probable to happen during residential fires or accidental breaks [35–37]. The potential for CdTe emissions could occur only in the case of industrial fires or from incinerating spent PV modules. In the first case, the fire itself probably would cause much greater risk than any potential Cd emissions [38]. The second case can happen only if CdTe modules end in waste-incineration streams [32].

4.2 Amorphous silicon (a-Si)

Amorphous silicon based solar cells are usually fabricated using the plasma enhanced chemical vapor deposition (PECVD) technique. Silane gas (SiH_4), mainly used as precursor, is extremely pyrophoric and represents the main safety hazard of this technology. It can spontaneously ignite for lower concentrations ranging from 2 to 3%, depending on the carrier gas. Due to the high pyrophoric nature of silane and even for concentrations lower than 2% in the carrier gas, pyrophoric footprints can be found locally if mixtures are not complete. Mixtures could be metastable and ignited after a certain time, for silane concentrations greater than 4.5% [32].

Amorphous silicon solar cells contain a large concentration of hydrogen atoms about 10%, as they are crucial for the material electronic properties [3]. But, the technology usually refers to use the words "amorphous silicon" instead of "hydrogenated amorphous silicon (a-Si:H)," because "unhydrogenated amorphous silicon" is of no use in electronic devices [3]. Hydrogen used in amorphous silicon manufacturing is explosive and flammable [32]; therefore, it is necessary for PV manufacturers to use highly sophisticated gas handling systems to minimize and even avoid the risks of fire and explosions. One efficient way to overcome these hazards is to store silane and hydrogen gases in bulk from tube trailers to avoid changing gas cylinders. Others toxic gases such as arsine (AsH_3), phosphine (PH_3) and germane (GeH_4), used as doping-gases in the amorphous silicon manufacturing cannot pose any serious hazards to the public health or the environment if they are used in very small amounts. However, leakage of these gases should be avoided because it could cause significant occupational risks.

4.3 Copper indium (gallium) diselenide (CI(G)S)

CIGS thin films can either be deposited by the thermal co-evaporation of the constitutive elements, or by the fast deposition of metal precursor layers which then react in a subsequent processing step to form the final compound [3]. In CIGS TFSCs, a very thin film of cadmium sulfide (CdS) is deposited by chemical bath method and acting as a buffer layer. However, CIGS solar cells freed of toxic cadmium have already been successfully produced [39]. The toxicity of copper, indium, gallium, and selenium is considered benign. In addition, elemental selenium is capital in the human nutrition; daily absorptions of 500–860 μg of selenium are acceptable for long periods [40]. Although elemental selenium has only a moderated toxicity associated with it, hydrogen selenide (H_2Se) used in the manufacture of CIGS TFSCs is highly toxic and is dangerous to life and health [32]. Hydrogen selenide acts like arsine gas on human body even though its vapor pressure is lower than that of arsine. Moreover, it can oxidize to the less toxic selenium on the mucous membranes of the breathing system. The manufacturing system should be enclosed under negative pressure, and should be exhausted through an essential control scrubber to prevent hazards from highly toxic H_2Se gas. Associated hazardous chemicals can be minimized by using safer alternatives methods like flow restricting valves and other safety options presented in detail by Fthenakis [41]. Some studies have shown that CIS and CGS have mild systemic toxicity and have shown no effects on ovulation, reproduction, liver and kidney [8]. But CIS was found to be less toxic than CGS and CdTe [8].

5. Methods to prevent environmental concerns

The releases of chemicals in form of vapors or spills from the PV industry constitute the real hazards to the public health and to the environment. A variety of treatment methods or ways have been developed to manage or to minimize wastes produced by PV industries. These methods include waste minimization and recycling of PV modules at the end of their life.

5.1 Waste minimization

Waste minimization is usually employed in the semiconductor industry, and is also appropriate to the PV sector. It includes reuse of rinse water after treatment, shifting toward less toxic chemicals as possible, control of spills and leaks, reduction of vapor losses, and selection of process that use fewer hazardous chemicals [8]. Some of these methods are not broadly used in the semiconductor sector due to the requirement of highly purified materials. The main goal to reuse processed chemicals in the semiconductor sector is to limit amounts of some harmful solvents and acids. A lot of changes have been successfully made in the manufacturing processes such as using less toxic materials instead of hazardous and replacing acid bath processing by acid spray in cleanings [33]. Splitting of spent solvents by type like chlorinated and non-chlorinated solvents can help increasing the volume of solvents that can be easily recycled [8].

Another way to minimize wastes generated during the fabrication of thin film solar cells and modules is reducing the amounts of toxic elements. For example, the possibility of reducing the quantity of toxic cadmium in the synthesis of CdS thin films, which plays the role of the buffer layer in CdTe and CIS solar cells has been investigated. It was found that by varying the solution concentration and temperature in the chemical bath deposition process for instance, the typical concentration of cadmium can be reduced up to 10 times [42]. These different ways to operate can help reducing substantially the amount of wastes generated during the PV manufacturing processes.

5.2 Recycling

It will be many years before most PV panels come to the end of their life (about 30 years), so it is needed to put in place some recycling schemes to prevent in time the harmful effects of spent panels on the environment. Some major PV manufacturers have experienced a promising approach called "cradle to cradle" [8]. The concept of this approach is to recycle the toxic materials of the process into new products, which are less or not at all toxic. This approach has enabled reducing the potential for release wastes into the environment, and enhancing the amount of new resources that must be obtained.

There are different ways or models of recycling PV modules implemented by PV manufacturers. Deficient PV systems from manufacturing plants and spent PV modules are collected by manufacturers for being recycled. The first intent was to apply the electronics model of recycling, which involves an intermediate company that would gather the spent PV modules, dismount them, and deal the usable parts. Unfortunately, this model is less suitable to the PV sector, because usable materials are very thin, and therefore, the modules are difficult to dismantle. An efficient way for recycling PV modules is to use large metal smelters to melt scrap PV modules. For example, save cadmium from CIS modules would need the use of a copper and zinc smelter, whereas CdTe cannot be melted in a zinc smelter as cadmium is a by-product of zinc mining. A method of recycling CdTe modules and developed by Solar Cells Inc. involves dismantling of the module, followed by glass milling and separation of the metals following a combination of physical and chemical methods such as chemical dissolution, mechanical separation, precipitation, and electrodeposition [43]. By this way, about 80% of the original tellurium was saved. Another method for recycling CIS and CdTe modules, and developed by Drinkaard Metalox Inc. uses chemical stripping, electrodeposition, precipitation, and evaporation. About 95% of tellurium and 96% of the lead for cells connection were saved by combining these different methods. This method allows the potential reuse of the substrate, because the metal conducting layer remains connected to the glass substrate after the separation of the elements [44]. A method of recycling CIS and CdTe modules based on electrochemical reactions in a closed loop system has been experienced at the early of years 2000 By Menezes et al. [45]. This approach could also lead to improve efficiency in the original CIS solar cells.

In addition to the environment safety, another major reason for developing relevant and cost-effective methods for recycling PV modules is the scarcity of some of the exotic elements used in PV industry. Reserves of some elements like germanium, indium and tellurium are low and continue to decrease with time [8]. Considering that the TFPV technology is still growing, it is needed to develop more feasible ways to recycle PV materials in order to preserve their reserves in the earth crust.

6. Conclusions

This chapter has shown the potential of some materials and chemicals used in the manufacture of thin film PV solar cells and modules to be hazardous. These hazardous chemicals can pose serious health and environment concerns, if proper cautions are not taken. Hazards could arise first from the toxicity and explosiveness of specific gases, then could affect occupational health and, in some cases, public health through accidents or elusive air emissions. Accidental releases of toxic gases and vapors can be prevented by minimizing wastes produced during the processes through choosing safer technologies, processes and less toxic materials. Recycling is expected to be the preferred disposal option for spent PV modules in the future, in order to minimize the potential environmental impacts and recover source of metals. Research is ongoing to build feasible methods of recycling spent modules for environmental safety.

Acknowledgements

The authors acknowledge ANSOLE (African Network for Solar Energy) for financial support.

Conflict of interest

The authors declare that they have no conflict of interest.

Notes/thanks/other declarations

Hervé Joël Tchognia Nkuissi is still grateful to the ICTP (The Abdus Salam International Centre for Theoretical Physics) and ANSOLE (African Network for Solar Energy) for financial support within the framework of the Intra-African Exchange (INEX) program, which helped him to complete his PhD studies at Hassan II University of Casablanca in Morocco.

Author details

Hervé Joël Tchognia Nkuissi[1,2]*, Fransisco Kouadio Konan[2,3], Bouchaib Hartiti[2] and Jean-Marie Ndjaka[1]

1 Laboratory of Mechanics, Materials and Structure, Department of Physics, Faculty of Science, University of Yaoundé I, Yaoundé, Cameroon

2 ERDyS Laboratory, Materials, Energy, Water, Modeling and Sustainable Development Group, Faculty of Science and Technique, Hassan II University of Casablanca, Mohammedia, Morocco

3 Laboratory of Solar Energy and Nanotechnology, Research Institute on New Energies, Nangui Abrogoua University, Abidjan, Côte d'Ivoire

*Address all correspondence to: hervetchognia@gmail.com

References

[1] Bain R, Denholm P, Heath G, Mai T, Tegen S. Biopower technologies. In: Hand MM, Baldwin S, DeMeo E, Reilly JM, Mai T, Arent D, et al., editors. Renewable Electricity Futures Study. Vol. 2. Golden, Colorado, United States: National Renewable Energy Laboratory; 2012. pp. 6-1-6-58. NREL/TP-6A20-52409-2.Ch6

[2] National Renewable Energy Laboratory (NREL). Best Research-Cell Efficiencies [Internet]. 2018. Available from: https://www.nrel.gov/pv/cell-efficiency.html [Accessed: February 11, 2019]

[3] Minnaert, B. Thin Film Solar Cells: An Overview [Internet]. 2008. Available from: https://biblio.ugent.be/publication/4238935/file/4238983.pdf [Accessed: February 12, 2019]

[4] Partain LD. Solar Cells and Their Applications. 1st ed. New York, United States: John Wiley and Sons Ltd; 1995. p. 600. ISBN: 9780471574200

[5] Goetzberger A, Hebling C. Photovoltaic materials, past, present, future. Solar Energy Materials and Solar Cells. 2000;**62**:1-19

[6] Haug FJ, Zurich ETH. Solar Cells: Thin Film Solar Cells [Internet]. 2018. Available from: http://www.tfp.ethz.ch/Lectures/pv/thin-film.pdf [Accessed: February 15, 2019]

[7] Andersson BA, Jacobsson S. Monitoring and assessing technology choice: The case of solar cells. Energy Policy. 2000;**28**:1037-1049

[8] Summers K, Radde J. Potential Health and Environmental Impacts Associated with the Manufacture and Use of Photovoltaic Cells. Vol. 98. Sacramento, CA, United States: EPRI and California Energy Commission; 2003

[9] Neumann H. Optical properties and electronic band structure of CuInSe2. Solar Cells. 1986;**16**(1-4):317-333. DOI: 10.1016/0379-6787(86)90092-X

[10] Moller HJ. Semiconductors for Solar Cells. ARTECH House; 1993. p. 343. ISBN: 978-0890065747

[11] Goetzberger A, Hebling C, Schock HW. Photovoltaic materials, history, status, and outlook. Materials Science and Engineering. 2003;**40**:1-46

[12] Spectrolab. Triple-Junction Terrestrial Concentrator Solar Cells [Internet]. 2003. Available from: http://www.spacedaily.com/reports/Spectrolab_Terrestrial_Solar_Cell_Concentrator_Maxs_Energy_Conversion.html [Accessed: February 16, 2019]

[13] Green MA, Hishikawa Y, Dunlop ED, Levi DH, Hohl-Ebinger J, Ho-Baillie AWY. Solar cell efficiency tables (version 51). Progress in Photovoltaics: Research and Applications. 2018;**26**: 3-12. DOI: 10.1002/pip.2978

[14] Yoshikawa K, Kawasaki H, Yoshida W, Irie T, Konishi K, Nakano K, et al. Silicon heterojunction solar cell with interdigitated back contacts for a photoconversion efficiency over 26%. Nature Energy. 2017;**2**(17032):1-5

[15] Benick J, Richter A, Müller R, Hauser H, Feldman F, Krenckel P, et al. High-efficiency n-type HP mc silicon solar cells. IEEE Journal of Photovoltaics. 2017;7:1171-1175. DOI: 10.1109/JPHOTOV.2017.2714139

[16] Verlinden PJ. Will we have > 22% efficient multi-crystalline silicon solar cells? In: Presented at PVSEC 26; Singapore. 2016. pp. 24-28

[17] Kato T, Handa A, Yagioka T, Matsuura T, Yamamoto K, Higashi S, et al. Enhanced efficiency of Cd-free

Cu(In,Ga)(Se,S)2 minimodule via (Zn,Mg)O second buffer layer and alkali post treatment. In: Proceedings of the 44th IEEE Photovoltaic Specialists Conference; 25-30 June 2017. Washington DC, New York: IEEE; 2017

[18] Sugimoto H. High efficiency and large volume production of CIS-based modules. In: 40th IEEE Photovoltaic Specialists Conference, Denver, 8-13 June 2014. 2014. pp. 2767-2770. DOI: 10.1109/PVSC.2014.6925503

[19] First Solar Press Release. First Solar Builds the Highest Efficiency Thin Film PV Cell on Record [Internet]. 2014. Available from: https://investor.firstsolar.com/news/press-release-details/2014/First-Solar-Builds-the-Highest-Efficiency-Thin-Film-PV-Cell-on-Record/default.aspx [Accessed: April 13, 2019]

[20] First Solar Press Release. First Solar Achieves World Record 18.6% Thin Film Module Conversion Efficiency [Internet]. 2015. Available from: https://investor.firstsolar.com/news/press-release-details/2015/First-Solar-Achieves-World-Record-186--Thin-Film-Module-Conversion-Efficiency/default.aspx [Accessed: April 13, 2019]

[21] Keevers MJ, Young TL, Schubert U, Green MA. 10% efficient CSG minimodules. In: Proceedings of the 22nd European Photovoltaic Solar Energy Conference; 3-7 September 2007. Milan; 2007. pp. 1783-1790

[22] Matsui T, Sai H, Suezaki T, Matsumoto M, Saito K, Yoshida I, et al. Development of highly stable and efficient amorphous silicon based solar cells. In: Proceedings of the 28th European Photovoltaic Solar Energy Conference; September 30–October 4 2013. Paris; 2013. pp. 2213-2217. DOI: 10.4229/28thEUPVSEC2013-3DO.7.2

[23] Kayes BM, Nie H, Twist R, Spruytte SG, Reinhardt F, Kizilyalli IC, et al. 27.6% conversion efficiency, a new record for single-junction solar cells under 1 sun illumination. In: Proceedings of the 37th IEEE Photovoltaic Specialists Conference; 19-24 June 2011; Seattle. New York: IEEE; 2012. DOI: 10.1109/PVSC.2011.6185831

[24] Mattos LS, Scully SR, Syfu M, Olson E, Yang L, Ling C, et al. New module efficiency record: 23.5% under 1-sun illumination using thin-film single-junction GaAs solar cells. In: Proceedings of the 38th IEEE Photovoltaic Specialists Conference; 3-8 June 2012; Austin. New York: IEEE; 2012. DOI: 10.1109/PVSC.2012.6318255

[25] Sun K, Yan C, Liu F, Huang J, Zhou F, Stride JA, et al. Over 9% efficient kesterite Cu_2ZnSnS_4 solar cell fabricated by using Zn1-xCdxS buffer layer. Advanced Energy Materials. 2016;**6**(12):1600046. DOI: 10.1002/aenm.201600046

[26] Fraunhofer Institute for Solar Energy Systems ISE. Photovoltaics Report [Internet]. 2019. Available from: https://www.ise.fraunhofer.de/content/dam/ise/de/documents/publications/studies/Photovoltaics-Report.pdf [Accessed: April 13, 2019]

[27] Azom. Photovoltaic – Concentrator Systems [Internet]. 2003. Available from: http://www.azom.com/details.asp?ArticleID=1173 [Accessed: February 13, 2019]

[28] Surek T. Photovoltaics: Energy for the New Millenium [Internet]. 2003. Available from: http://www.aps.org/units/fps/meetings/april2000/surek.html [Accessed: February 13, 2019]

[29] Powalla M, Bonnet D. Thin-film solar cells based on the polycrystalline compound semiconductors CIS and CdTe. Advances in OptoElectronics. 2007;**2007**:6. DOI: 10.1155/2007/97545

[30] Hill R, Baumann AE. Environmental costs of photovoltaics. IEE Proceedings A Science, Measurement and Technology. 1993;**140**:76-80. DOI: 10.1049/ip-a-3.1993.0013

[31] Moskowitz PD. An overview of environmental, health and safety issues in the photovoltaic industry. In: Partain LD, editor. Solar Cells and Their Applications. 1st ed. New York, NY: John Wiley & Sons, Inc.; 1995. pp. 391-416. Chapter 18

[32] Fthenakis VM. Overview of potential hazards. In: Markvart T, Castaner L, editors. Practical Handbook of Photovoltaics: Fundamentals and Applications. Elsevier; 2003. pp. 1-14. ISBN 1-856-17390-9.chVII-2

[33] Gilles DG, Loehr RC. Waste generation and minimization in semiconductor industry. Journal of Environmental Engineering (ASCE). 1994;**120**:72-86

[34] Drysdale D. An Introduction to Fire Dynamics. 1st ed. Wiley; 1985. p. 424

[35] Patterson M, Turner A, Sadeghi M, Marshall R. Health, safety and environmental aspects of the production and use of CdTe thin film modules. In: Proceedings of the 12th European PV Solar Energy Conference; 11-15 April 1994; Amsterdam. 1994. pp. 951-953. ISBN: 0952145235

[36] Thumm W, Finke A, Nuemeier B, Beck B, Kettrup A, Steinberger H, et al. Environmental and health aspects of CIS-module production, use and disposal. In: Proceedings of 1994 IEEE First World Conference on Photovoltaic Energy Conversion; 5-9 December 1994; Hawaii. New York: IEEE; 1994. pp. 262-265

[37] Steinberger H. HSE for CdTe and CIS thin film module operation. In: Niewlaar E, Alsema E, editors. IEA Expert Workshop on Environmental Aspects of PV Power Systems; May 23, 1997, Report No. 97072. The Netherlands: Utrecht University; 1997

[38] Bohland J, Smigielski K. First solar's CdTe module manufacturing experience; environmental, health and safety results. In: Proceedings of the 28th IEEE Photovoltaic Specialists Conference; 17-22 September 2000; Anchorage, Alaska. New York: IEEE; 2000. pp. 575-578

[39] Miles RW, Hynes KM, Forbes I. Photovoltaic solar cells: An overview of state-of-the-art cell development and environmental issues. Progress in Crystal Growth and Characterization of Materials. 2005;**1**:1-42

[40] Piscator M. The essentiality and toxicity of selenium. In: Carapella SC, editor. In: Proceedings of the Fourth International Symposium on Uses of Selenium and Tellurium; 7-10 May 1989; Alberta, Canada

[41] Fthenakis V. Multi-layer protection analysis for photovoltaic manufacturing. Process Safety Progress. 2001;**20**(2):1-8

[42] Bayer A, Boyle DS, Heinrich MR, O'Brien P, Otway DJ, Robbe O. Developing environmentally benign routes for semiconductor synthesis: Improved approaches to the solution deposition of cadmium sulfide for solar cell applications. Green Chemistry. 2000;**2**:79-86. DOI: 10.1039/A909257I

[43] Bohland J, Dapkus T, Kamm K, Smigielski K. Photovoltaics as hazardous materials: The recycling solution. In: Fthenakis V, Zweibel K, Moskowitz P, editors. Presented at the BNL/NREL Workshop on Photovoltaics and the Environment 1998, Keystone, Colorado, July 23-24; 1998. pp. 33-40

[44] Goozner R, Byrd C, Long M, Drinkard WF. Recycling thin film photovoltaic materials. In: Fthenakis V, Zweibel K, Moskowitz P, editors. Presented at the BNL/NREL Workshop

on Photovoltaics and the Environment 1998, Keystone, Colorado, July 23-24. 1998. pp. 42-59

[45] Menezes S. Electrochemical solutions to some thin-film PV manufacturing issues. Thin Solid Films. 2000;**361-362**:278-282

Chapter 8

Emerging Photovoltaic Technologies and Eco-Design—Criticisms and Potential Improvements

Nicole Mariotti, Matteo Bonomo and Claudia Barolo

Abstract

An efficient waste management for emerging photovoltaic (PV) technologies is not mature yet. The problematic aspects along with the possible failure's identification have a pivotal role in modelling the future end-of-life management strategies. The identification of substances of concern (e.g. high cost, low availability, and high toxicity) and valuable materials is a key point to better define the research priorities to improve the eco-design of these technologies. The ultimate goal is to promote the disposal processes which enhance the repair, refurbishment, and recover opportunities and so the profitability of recycling. These studies can also prompt the investigation of innovative materials which are more cost-effective and/or coming from renewable resources or secondary raw materials. Forecasting the waste management technologies for the emerging photovoltaics is highly challenging. In this context, our purpose is to provide an overview of the critical elements and understand the appropriate corrective improvements towards more sustainable technologies.

Keywords: sustainability, environmental impact, emerging photovoltaic technologies, life cycle assessment, recycling, disposal, organic photovoltaics, dye-sensitised solar cells, perovskite solar cells

1. Introduction

We must recognise that overpopulation is the main environmental issue, the main cause of the unsustainable depletion of our planet. The correlation between overpopulation and environmental concerns can be completely understood through the definition by Ehrlich et al. [1].

When is an area overpopulated? When its population cannot be maintained without rapidly depleting non-renewable resources (or converting renewable resources into non-renewable ones) and without decreasing the capacity of the environment to support the population. In short, if the long-term carrying capacity of an area is clearly being degraded by its current human occupants, that area is overpopulated.

Recently, some studies [2, 3] confirmed that human population planning (anti-natalist policies) cannot stop the enormous environmental crisis that we are facing nowadays. Indeed, a prompt and effective solution should be found in technology and through the improvement of resources and waste management.

Energy production is for sure at the top of the list of the most polluting activities in the whole planet. For this reason, environmental concerns are strengthening the interest in the alternative energy research field that started already during the 1970s energy crisis. The energy transition should intimately deal with the implementation of new infrastructures, the promulgation of adequate policies and the gradual conversion of a system that has been based on oil for more than 150 years [4].

In the context of energy production, one of the most promising alternative sources to fossil fuels is solar energy. Indeed, the energy supplied by the sun irradiation over 1 year is roughly 10,000 times higher than the world's current rate of energy consumption [5]. Starting from the first photovoltaic (PV) technologies based on silicon, the researchers developed a plethora of different materials and devices looking for higher efficiencies and performances. Nevertheless, the design of a new technology cannot be set up only on efficiency and performance improvements but also the sustainability of the product must be considered. Starting from this assumption, it becomes more and more important to examine the following aspects when a new technology is studied: (i) the energy consumed for its production, (ii) the kind of resources and materials exploited, (iii) the waste produced during the production and use phases, (iv) the energy produced/energy consumed ratio and (v) the end of life of the product. It is worth mentioning that these aspects are not only strictly related to environmental issues but also to the high dependency of countries on resources, the critical problem of waste accumulation and the increasing cost of waste disposal—these are also economic and sociologic issues to be seriously addressed. The increasing demand for raw materials and the waste accumulation can be partly mitigated through the recycling and the identification of secondary raw materials. This is particularly important for those materials defined as Critical Raw Materials (CRMs), that is, materials that are characterised by supply risk and economic importance (**Figure 1**).

A very powerful tool to analyse how sustainable is a product or a service is the *Life Cycle Assessment* (LCA); this helps to evaluate the environmental impacts of a

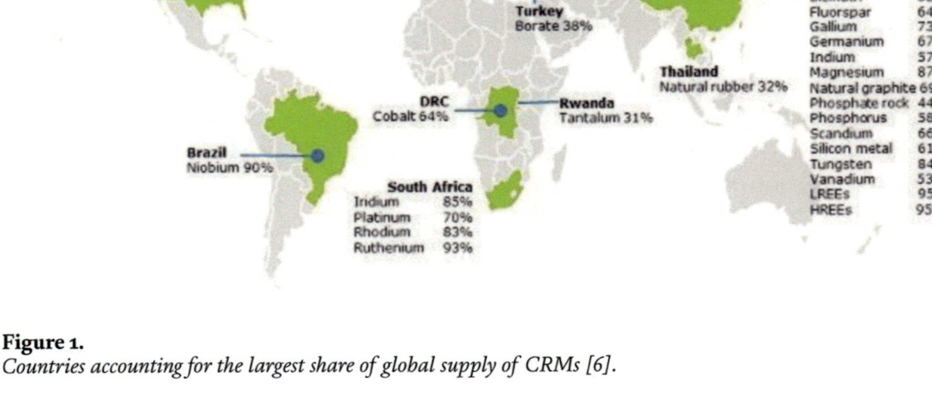

Figure 1.
Countries accounting for the largest share of global supply of CRMs [6].

product life-cycle in all its parts. In early technology development stage, an LCA can give an overestimation of the energy consumed; it is important to underline that laboratory instruments are not completely conceived for a specific synthesis or process. However, it could be of great relevance—it has been indeed recognised that 70% of a product's environmental impact is determined in the design and development stage [7, 8]. The life cycle assessment can have either a cradle-to-gate or a cradle-to-grave approach. The former allows to outline the environmental profile of a product, process or service from the raw materials extraction to the factory gate (the use phase and disposal phase of the product are omitted in this case); the latter considers the whole process, from the raw materials extraction to the disposal phase [9].

In this chapter, many LCA works have been analysed to assess the sustainability of emerging PV technologies. There are many variables in life cycle assessment; any study can consider a different functional unit, a different efficiency and a different life-time of a device. Hence, it is recommended to check the referenced LCA works to have a better comprehension of the boundaries and limitations of the studies as it was not possible to give all the details in the body of this work.

2. Emerging PV technologies—critical aspects and potential improvements

Although the energy generated by silicon PVs is quickly approaching, the price of traditional sources (such as coal) and the market prices of silicon are decreasing (0.30 $ W/peak [10]), the interest on emerging PVs is still present. These technologies exhibit a plethora of advantages: lower materials cost, easy manufacturing, flexible and light modules, less energy and mass requirement in the production with a subsequent decrease of the overall environmental footprint.

In the field of solar energy conversion, emerging technologies include Organic Photovoltaics (OPVs), Perovskite Solar Cells (PSCs) and Dye-Sensitised Solar Cells (DSSCs). In the following sections, an analysis of the critical aspects of each class of devices will be presented in order to highlight weaknesses and failures and to identify the corrective measures to apply to build more sustainable PVs. Disposal perspectives will be investigated as an efficient waste management is not mature yet. In this context, it is worth mentioning that the studies considered hereafter mainly concerns laboratory-scale device. Yet, the strategies highlighted could be feasibly extended on larger scale too.

2.1 Organic photovoltaics

Organic photovoltaics are a promising technology as it offers the advantage to have light, easily printable and scalable, thin and flexible solar panels [11].

Organic solar panels are usually structured as follows:

1. A flexible polyethylene terephthalate (PET) substrate on the light collecting side;

2. A transparent conductive oxide (TCO), usually indium tin oxide (ITO) or fluorine-doped tin oxide (FTO), as electrode

3. An active layer usually made of a [6,6]-phenyl-C61-butyric acid methyl ester:poly-3-hexyl-tiophene (PCBM:P3HT) mixture behaving as electron transport layer (ETL)

4. A hole transport layer (HTM) usually based on a PEDOT:PSS system but also MoO_3 can be used

5. An aluminium back electrode covered by a thin layer of lithium fluoride

6. Sometimes, a second PET layer could be employed to cover (and straightforwardly protect) the entire device.

7. Epoxy resins as sealant material

8. Interconnections made of a silver paste

The configuration of the devices can be of two different types: the single- and multi-junction system. The former is a single p-n junction while the latter has more than one p-n junction leading to better efficiencies but requiring a higher amount of materials coupled to more elaborated production processes (**Figure 2**).

OPVs usually exhibit lower efficiency and shorter lifetime compared to silicon panels, but they always show lower cradle-to-gate life-cycle impacts [11]; OPV panels can save around one-sixth and one-fourth of the cradle-to-gate energy consumption compared to m-Si and a-Si (monocrystalline and amorphous silicon) panels, respectively [12].

Even if OPVs show higher environmental efficiency with respect to the more diffused silicon technologies, some failures have been determined.

PET, that is used both as a substrate and encapsulant, is a non-biodegradable polymer under environmental conditions. Even if it is not considered to be directly toxic, PET can be dangerous for the environment and can show ecotoxic issues when it is broken into small particles (diameter smaller than 5 mm) [13]. If so finely dispersed, it can reach organisms through the food chain. It has also been demonstrated that PET particles can sorb other persistent organic pollutants and, straightforwardly, these compounds can reach organisms more easily. Photo-degradation is another critical parameter considered in the literature [14]: the natural photo-degradation of PET leads to the formation of both monomer and dimer fragments. Thus, the employment of a stable substrate and enclosing system are essential to have a reasonable lifespan. Bio-plastics (such as PLA) have been considered to be adopted instead of PET, but even if they present a better LCA, their mechanic stability is not good enough. Straightforwardly, it is compulsory to design a configuration such as the persistent materials (not bio-degradable) can be easily recovered and then recycled [14]. In particular, PET recovery can be performed only through an efficient delamination process. In this context, it is necessary to identify the best encapsulant material that can be thermally or chemically separated from

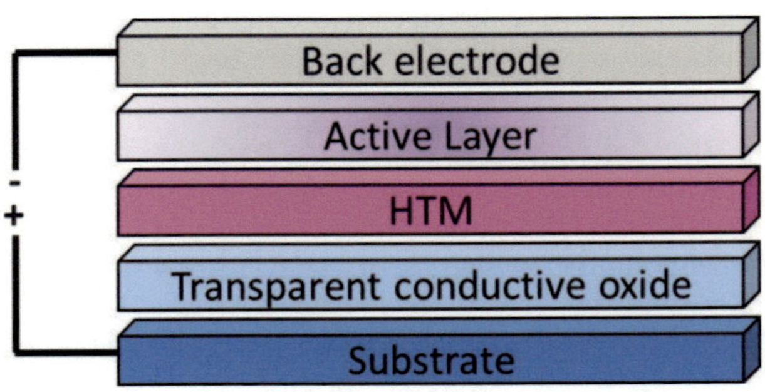

Figure 2.
General structure of an OPV device.

the substrate. It should have a different solubility or a melting point lower than the other device components. As far as we are aware, nowadays, researchers are focused on various features of an encapsulant such as chemical inertness, water vapour transmission rate (WVTR) and oxygen transmission rate (OTR), but recycling or recover perspective are not considered yet [15].

ITO, applied as conductive coating on the PET substrate, tends to etch by air and water becoming then a pneumo-toxic entity and a strong reactive oxygen species (ROS) producer when it is in its nanoparticulate form [16]. Further studies are still necessary to understand the fate aspects of ITO in the environment, but its behaviour should be more carefully investigated to highlight potential criticisms. The presence of indium, that is a CRM lead to focus the attention on the recovery of this material. Alternatively, efforts should be made on the research of similar but more environmental friendly, for example, FTO [14]. Nevertheless, FTO is defined as a hotspot in a cradle-to-grave OPV LCA too [11].

A meaningful focus on the active layers has been carried by different studies [14, 17, 18].

Zimmermann et al. [14] focused their attention on the degradation and the ecotoxicity of PCBM/P3HT active layer. The results show that P3HT is degraded by a radical building process. The formed compounds can be oxidised, and this reaction is found to be responsible for the degradation of the organic polymer layer. Although ecotoxicity data for P3HT are not available, they suggested to investigate the potential degradation in small pieces similarly to what happens with PET. The second component of the active layer is the fullerene derivative PCBM. Fullerenes' toxic effects have been thoroughly discussed in the literature [19, 20]: they exhibit acute toxicity and can sorb pollutants making them more bioavailable. These studies focused on fullerenes and not on their derivatives, such as PCBM. Therefore, further critical analyses on these compounds are essential to assess their fate and their ecotoxicity [14]. Very interestingly, Tsang et al. [11] reported a comparison between the OPV based on PCBM/P3HT and completely polymeric OPV (i.e. changing PCBM with an n-type polymer) showing that the latter is slightly better for some impact factors than the former one. In particular, an all-polymer-based device exhibits a water depletion factor lower (−12%) compared to the counterpart based on PCBM/P3HT. The substitution of the fullerene derivative with an n-type polymer has been suggested from the results of previous studies that identified PCBM as an important factor in the increase of the Cumulative Energy Demand (CED) component [17, 21–24]. CED is defined as an energetic indicator that quantifies the whole energy required during the life cycle of a product. It is obtained summing up both the direct energy (e.g. electricity) and the indirect energy (embodied energy of materials) contributions [25].

The investigation of 15 different materials implemented as active materials in OPVs showed that fullerenes and their derivatives (used as electron-acceptors) have the highest CED in a cradle-to-gate LCA [17]. On the other hand, electron donor materials, such as polymers or small molecules, affect less sensibly the total cradle-to-gate CED. A deeper analysis on fullerenes showed that functionalized ones have a higher CED than the native counterparts, while the indene-C60 bis-adduct (ICBA) derivatives present a CED 40% lower than C60-PCBM and C70-PCBM. This behaviour can be explained by the fact that ICBA derivatives have a simpler reaction scheme that results in a lower number of synthetic step; in this way, the purification steps and the use of solvents are decreased. Fullerenes in any case have an embodied energy one or two orders of magnitude higher than the common semiconducting polymers [17]. The semiconducting P3HT is the polymer with the lowest CED because the number of steps for its synthesis is relatively lower; generally, the doubling of the number of steps lead to a 10-fold increase in the CED. Block copolymers, such as PCDTB and PTB7, have a higher embodied energy but they show a lower CED with respect to fullerenes.

Interestingly, phthalocyanines and squaraines can be used in place of electron-donor polymers and have been thoroughly examined [17]; comparing these two classes of molecules, they show very similar CED except for the PdPc that has a higher embodied energy due to the Pd extraction. These small molecules compered to polymers showed a similar CED.

The embodied energy of polymer-based devices is mainly due to the intrinsic energy of materials while for the small molecule-based cells to the processing conditions (i.e. fabrication step). For this reason, multi-junction small molecule-based devices are energetically convenient compared to mono-junction. Indeed, for small molecule-based devices, the process to build a multi-junction device requires almost the same energy compared to a mono-junction, and the increase in the amount of required material does not sensibly affect the CED.

The hole transporting layer (HTL), usually made of a composite of PEDOT:PSS or MoO_3, contributes in a very minimal way to the overall CED. In particular, Anctil et al. [17] reported that, concerning PEDOT:PSS, the solvent used for the deposition is the main energy increasing factor. The interfacial layer can be also responsible for the device degradation. PEDOT:PSS is an hygroscopic material, and it is quite sensitive to oxidation reactions; this could cause an extremely fast degradation of the organic panels. PEDOT materials, in their particulate form, showed cytotoxicity, ROS production, apoptosis, and necrosis [26]. Such as for ITO, particular attention must be paid to a potential dispersion in water.

Finally, considerations about the use of silver must be done. Silver is one of the main responsible for resource depletion and freshwater eutrophication [14]. Espinosa et al. [27] carried out a cradle-to-grave (from the materials supply to the end-of-life) LCA of an organic solar park. They compared three different end-of-life scenarios: recycling, incineration and an average local mix. The results showed that the recycling scenario is favoured mainly because of PET and silver recovery.

Silver should be recovered or replaced by a non-metal electrode and bio-based electrode materials or production from waste should be considered. Alternative materials such as carbon, copper and aluminium have been studied, and the environmental impacts were proved to be the lowest for carbon-based modules [28]. The recovery of silver is of great importance also to avoid soil contamination. As a matter of fact, when a damaged panel is in contact with soil, the silver release is highly enhanced [18].

If the balance of system (BOS), consisting of mounting-structure, inverter and cables for electrical installation, is considered into the LCA of OPVs, the results showed that BOS is what affects more the apparatus impacts. On the other hand, the a-Si impacts depend more on the contribution from the panel itself. This shows that the eco-design should be more focused on the BOS improvement in OPV technologies. The contribution of BOS allowed to state that, generally, if the efficiency of the device decreases, the impact is higher; this relation depends on the increasing contribution of the background system [11, 27].

Organic photovoltaics show many advantages not only for their functionalities but also for an environmental point of view. They generally show better environmental performances than Si-PV and also lower Energy PayBack Time (EPBT) and Carbon PayBack Time (CPBT) compared to m-Si.

The EPBT is expressed in years and quantifies the time that the system takes to generate the same amount of energy used in all production processes [9]. The CBPT is the ratio between the CO_2-equivalent emissions during the life-cycle of a PV panel and the CO_2-equivalent emissions of a conventional grid for the same kWh produced by the panel [11].

The factor that usually affects the impacts in a negative way is the too short lifetime of these devices. Thus, the increase in the lifetime of the panels is a feasible strategy to decrease their impact [11]. It is worth remembering that PET, PEDOT:PSS and P3HT are not fully biodegradable.

It could be interesting to re-think the OPV design in order to improve the recyclability and the biodegradability in environmental conditions of its components. The external apparatus should be resistant and durable during the use-phase and recyclable at its end of life; the internal materials that are often mixed or difficult to separate should be degradable when they reach the environment. Obviously, this is necessary if the risk of leaks is present, otherwise they can be collected and degraded in the suitable plants with adequate conditions.

The eco-design of a product is always a challenging perspective but it is what the paradigm that outlines the rules for a more sustainable economy dictates [29].

2.2 Dye-sensitised solar cells

Dye-sensitised solar cells (DSSCs) were invented by O'regan and Grätzel in 1991. A typical DSSC device consists of (**Figure 3**):

1. A transparent anode made of glass coated with a TCO (usually ITO or FTO)

2. A mesoporous TiO_2 layer

3. A monolayer of dye adsorbed onto TiO_2

4. An electrolyte

5. A glass cathode coated with a catalyst (usually Pt)

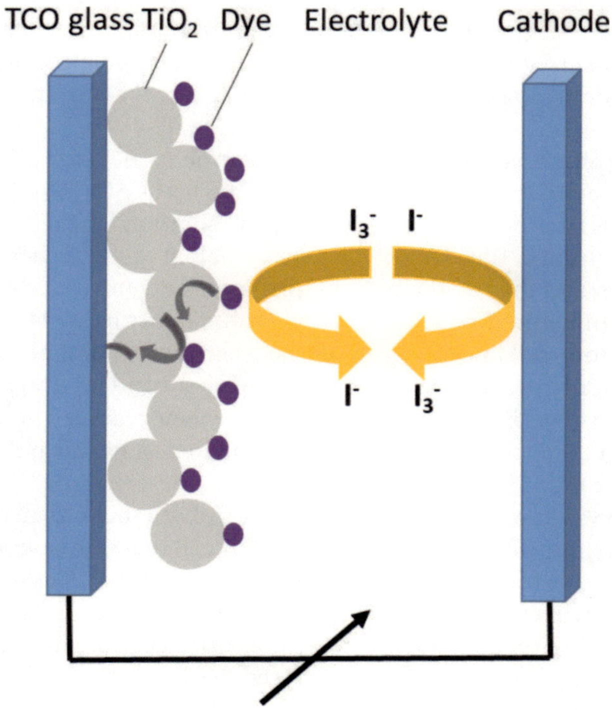

Figure 3.
Schematic diagram of the dye-sensitised solar cells.

When exposed to the sunlight, dye electrons are excited, and they are injected into the mesoporous material (titanium dioxide). Electrons move from the mesoporous layer to the anode and then into the circuit reaching the cathode. At this point, electrons are used to close the cycle: the electrolyte is reduced by the electrons coming from the circuit and in turn it re-reduces the dye that can be excited again [30].

The traditional electrolyte used in DSSCs is an organic solvent-based liquid electrolyte, usually iodide/triiodide (I^-/I_3^-) redox couple. It has been observed that this redox couple, usually dissolved in acetonitrile, is one of the main factors that affects the device lifetime [31]. In particular, the leakage of the solvent, evaporation of volatile iodine ions and corrosion are the dominant causes of damaging. In view of this fact, alternative redox couples [32–34] and different phases for the electrolytes have been investigated.

Among the alternative redox shuttles, the best performances have been obtained with cobalt complexes [31]. Cobalt is defined as a CRM in Europe: it is principally mined in the Democratic Republic of Congo that is a politically sensitive region. In 2017, the Financial Times reported the necessity to investigate the ethical issues regarding cobalt mining [35]; the establishment of the Cobalt Institute "promoting the sustainable and responsible use of cobalt in all forms" says a lot about the urgency to regulate the supply and use of this element. It should be clear, that this element, nowadays is mainly used in rechargeable batteries, is not a good candidate for a sustainable photovoltaic device.

Liquid-state electrolyte drawbacks can also be overcome through the implementation of quasi-solid (gel) and solid-state electrolytes [36]. Quasi-solid electrolytes are obtained by the entrapment of the liquid electrolyte into polymeric or inorganic networks. They present the best trade-off between efficiency and durability [37]. On the other hand, solid electrolytes exhibit a lower efficiency but the advantages of solid phase materials. Generally, solid materials are safer because they prevent leakage and evaporation; therefore, emissions and inhalation are avoided. Solid-state electrolytes also guarantee a better mechanical stability and simplified fabrication processes. In order to obtain such a geometry, the liquid electrolyte is replaced with an inorganic p-type semiconductor [38] or an organic hole transporting material [37].

Keeping the phase unchanged (i.e. liquid electrolytes), it is worth to mention both room temperature ionic liquids [39] and aqueous DSSCs [36]. Room temperature ionic liquids have been suggested mostly as a solution for the high volatility of organic solvents in liquid electrolytes. These molten salts are chemically and thermally stable, and they do not present inhalation and emission issues, due to the low vapour pressure (low volatility) and flammability hazards. Anyway, ionic liquids are controversial materials from a green chemistry point of view: on the one hand, they are really stable and safe, and on the other hand, they are not completely sustainable from a synthetic point of view and for their recovery [40, 41]. Very recently, deep eutectic solvents (DESs) have exploited as a valid and completely green alternative to ILs [42, 43].

Aqueous DSSCs have also been studied in order both to avoid all the negative effects of the organic solvents and to build a solar cell that is inherently thought to work in the presence of water moisture [36]. In fact, some studies [44–46] showed that water is a detrimental factor for DSSCs. Instead of performing laborious processes to avoid water penetration, researchers tried to build partially or fully aqueous solar cells [47, 48]. Water is considered to be the greenest solvent as it is safe, not-flammable and non-toxic but attention must be paid to a couple of aspects when water is considered. The first one is that fresh water is a limited resource

[49, 50]; as reported by Shiklomanov [49] and USGS [51], fresh water resources are just 2.5% of total water amount present in the world and only a small part is really available. This means that available water is less than the aforementioned 2.5%. The second significant point to stress is that any impurities or contaminants released in aqueous waste streams will, by their nature, readily find their way into aquifers, enhancing the risk of human exposure. Water is inherently a safe substance, but its precious value should be recognised. The transition from an organic solvent to water, especially when organic molecules are involved, has its own barriers. Initially, chemists have put their efforts either in the optimization of aqueous DSSCs using traditional electrolytes and dyes or in the investigation of not-fully aqueous (mixture of water and organic solvents) electrolytes [52]. Nowadays, the major trend is to re-think completely the chemistry aiming to completely aqueous DSSCs, using, for instance, surfactants [53]. It is still worth highlighting that water displays many attractive features as a solvent, but it is not without its own set of problems though.

LCA studies on dye-sensitised solar cells assessed that the main contributor to energy and environmental impacts is the coated glass [9, 54, 55] due to the high consumption of energy during the production and its preponderant presence with respect to the total mass.

Comparing three different dyes, Parisi et al. [9] found out that N719, a ruthenium-based dye is one of the main contributors to many cradle-to-gate impacts such as metal depletion, marine and freshwater ecotoxicity. The other two investigated sensitizers are D5, an organic metal-free dye and YD2-o-C8, which is a zinc-based porphyrin. The latter two dyes show lower impacts in all impact categories except for the ozone depletion parameter, mainly due to the massive use of solvents throughout their synthesis. As a further focus on these dyes, CEDs was carried out: N719 owns the lowest CED value compared with the others. This factor can be explained by the fact that the Ru-based dye (and straightforwardly its synthetic pathways) is the most optimised one. Anyway, for any dye, the total CED is mainly attributed to the embodied energy of raw materials and, to a minor extent, for the energy needed for the synthesis. Optimisation of synthetic processes for other kind of molecules can lead to better LCA results.

Even if ruthenium-based dyes require less energy and show lower impacts, attention on the scarce material ruthenium must be paid. In 2017, the European Commission defined ruthenium as a CRM as its supply is concentrated in South Africa and it presents low substitution and low recycling rates [37]: this statement strengthens the necessity of fully organic dyes to avoid the use of rare elements. The CEDs obtained from the LCA show that considering different configurations, the main component that affect the total energy is the embodied energy of raw materials and not the energy consumed during the production process of the device. This result lead to think that is more important to choose less energy demanding materials than improve the manufacturing process.

DSSCs (14.3%) [56] have not reached the efficiency of silicon-based photovoltaics yet. Anyway, their use in smart windows or their indoor application [57] gives the opportunity to this technology to enter the market, in particular in the building-integrated photovoltaics [58]. The added value of DSSCs is that they are not only functional but they also have a pleasant aesthetics.

Finally, it is worth highlighting that DSSCs can achieve further improvements in terms of sustainability through the implementation of nature-based dyes [48, 59], the use of alternative substrates [60], such as paper [61, 62] and the re-design of the devices in order to make them adequate for dye refurbishment. Obviously, these improvements should be coupled to give enough high photoconversion efficiency.

2.3 Perovskite solar cells

The name perovskite was firstly used for the calcium titanium oxide ($CaTiO_3$) mineral in 1839, named after the Russian mineralogist Lev Perovski. In 1957, Christian Møller discovered that caesium lead halides ($CsPbX_3$, X = Cl, Br or I) owned the same structure of $CaTiO_3$ [63] and found out that they were photoconductive behaving as semiconductors.

An important step towards the development of photovoltaics based on perovskites was made when Weber replaced caesium with methylammonium cations and obtaining in this way an organic-inorganic hybrid perovskites [64]. In particular, methylammonium lead iodide (general formula: $CH_3NH_3PbI_3$) is one of the most implemented materials in PSCs. It is a semiconducting pigment that can absorb light over the whole visible solar emission spectrum. Excitons produced in $CH_3NH_3PbI_3$ dissociates very rapidly into excitons (i.e. free carriers of different sign) and exhibits high carrier mobilities for electrons and holes and also long carrier-diffusion lengths when coupled with efficient Electron Transport Layer (HTL) and Hole Transport Material (HTM), respectively. These properties are responsible for the great potential of hybrid perovskites in photovoltaics. The choice of the different HTL and HTM should be thoughtfully made, because they should be inert towards the Perovskite layer but, on the other hand, they should assure a good electronic matching (**Figure 4**).

The first studies on perovskite solar cells (PSCs) were carried out only in 2009 [65] and since then different types of PSCs were developed as they are promising for their relatively low cost and high power conversion efficiency. These advantages could be eclipsed by the potential toxicity of lead even if many studies revealed that the main environmental impacts are due to the presence of gold (as back contact) and the use of organic solvents throughout the fabrication process [66–68]. For a better understanding of these impacts, a description of PSC modules and an overview of their hotspots will be given.

A perovskite solar cell is usually composed of (**Figure 4**):

1. A glass substrate

2. An ITO or FTO layer

3. A compact layer made of TiO_2 (SnO_2 and ZnO)

4. A mesoporous or compact TiO_2 layer as electron transporting layer (ETL) on the previous layer.

5. Perovskite ($CH_3NH_3PbI_3$) layer

6. A hole transport material (HTM) layer

7. A gold or silver cathode

The most implemented technologies are the mesoporous and the planar structures: the former involves the use as electron transporting material of a compact and a mesoporous layer of TiO_2, while the latter is composed only of a compact layer (**Figure 4**) of TiO_2, SnO_2 or other semiconducting metal oxides.

The main point to consider regarding the environmental impact of PSCs is that ITO or FTO glass represents about the 97% of the total mass of the modules; for this reason, the substrate is the main contributor to the energy consumption due to the embedded energy of materials.

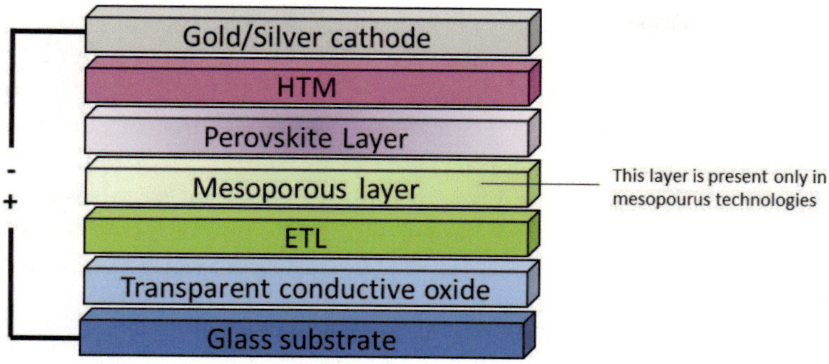

Figure 4.
General structure of a perovskite solar cell.

When gold is used as cathode, the energy consumption is mainly equally distributed between the substrate [ITO (FTO)-glass] and gold. The substitution of gold with silver can decrease not only the energy demand but also the environmental impact: gold is also mainly responsible for eutrophication, fresh water aquatic ecotoxicity, fresh water sediment ecotoxicity, human toxicity, land use, marine aquatic ecotoxicity, marine sediment ecotoxicity, depletion of abiotic resources, stratospheric ozone depletion, and terrestrial ecotoxicity [68]. Deposition of gold is also an important factor affecting the energy demand: its deposition and evaporation under vacuum is an intensive-energy process. Anyway, we have to consider all the issues belonging to silver already cited in the OPV section.

As we already said for the other technologies, ITO glass should be replaced with the FTO substrate since avoiding the use of precious metal (In) improves the environmental impacts of the modules.

Among the TCO/glass and the gold cathode, solvents used to rinse the ITO (FTO) glass and to prepare the perovskite layer play an important role in the environmental impacts. Even if they could be partially recycled, solvents show negative impacts due to the electricity needed in the recycling process; in particular, the solvent recycling exhibits the worst environmental performances for the ozone depletion potential and for the global warming potential [66]. Straightforwardly, the use of solvents should be directly downgraded or greener solvents should be preferred [69, 70].

Since the lead toxicity is of really high concern [71], many studies focused their attention on this element: the impacts of lead derivatives and disposal scenarios have been analysed in detail. Lead can cause irritability, difficulty in concentrating, headache, and anaemia and in higher concentrations (over 100 μg/dL) gives rise to seizures, delirium and even coma [72]. Lead toxicity is well documented, so regulation promoting the gradual phasing out of this substance has been promulgated by EU [73]. The Restriction of Hazardous Substances Directive (RoHS) [73] affirms that the maximum permitted concentrations are 0.1% or 1000 ppm by weight. The restrictions are on each homogeneous material in the product and understanding. If lead halide perovskites are considered homogeneous materials or not is a matter of study. If perovskites were considered homogeneous materials, they would not meet the regulatory restrictions. The regulation is not applied to PVs intended for a defined location (e.g. solar parks) but put limits on portable devices. In this case, where the regulation must be followed, PSCs would not match the 0.1% limit as lead is present in higher concentrations.

Moreover, lead presence could affect consumers' choices as the public opinion owns a solid awareness of lead toxicity raised through historical facts such as Romans poisoning caused by lead present in drinking-water, Goya's and Van Gogh's mental disorder probably due to lead contained in their colours and Kabwe inhabitants poisoned by mining activities.

In contrast to the high concern for lead toxicity, the results reported by Alberola-Borràs et al. [74], Billen et al. [75] and Zhang et al. [66] deserve particular attention. They have demonstrated that the content of Pb in the perovskite contributes to the human toxicity cancer impact category 1–2 orders of magnitude less than the rest of the module components [Glass/ITO (FTO), ETM, HTM and back contact] [74]. Furthermore, the energy used for the manufacturing of panels is the dominant contributor to lead emissions [75]. Zhang et al. highlighted that lead contributes less than 1% to the Human Toxicity Potential (HTP) and Ecotoxicity Potential (ETP). It is possible to conclude that lead is not the factor of main concern for these parameters although the widespread belief.

Commercialization of PSCs is still far to be achieved. This technology presents short life-time, substantial stability issues and lacks scale-up processes and waste disposal strategies. In order to overcome these limits, researchers foresaw and analysed different emissions and disposal scenarios. Alberola-Borràs et al. [74] studied three disposal scenarios for both planar and mesoporous structures. The three considered disposal routes were: (i) inertization and residual landfill, (ii) re-use and residual landfill and (iii) reuse and recycling. For re-use, they mean regenerative cycles suggested from Kim et al. and Huang et al. [76, 77]. As largely expected, reuse strategies reduced the values of all impact category. Additionally, they found that the device with a mesoporous TiO_2 substrate is the one with the largest improvement in all impact categories. This highlights that the mesoporous TiO_2 reuse is of great importance to reduce the environmental impacts, while Pb derivatives recovery does not affect the performances in a significant way.

Billen et al. [75] considered potential lead emissions during the use phase and two different end-of-life scenarios. They proved that the lead emissions are dominated from those produced from the energy production for the manufacturing of the panel itself and of the BoS; indeed, the components that do not include the panel (BoS) are the main contributors. This study analysed a disposal scenario in which no lead release was supposed, and a second case in which lead is completely released in groundwater. They demonstrated that even in the worst case, the toxicity potential is smaller than that one calculated for an offset grid [75].

Although lead toxicity potential concern is minimised by different researches, aforementioned studies highlight the necessity to avoid incidental release, adopting proper encapsulating materials and maintenance measures, as well adequate disposal strategies. Encapsulant should be resistant and stable under environmental conditions but easily removable through solvent dissolution or by thermal decomposition without the production of toxic substances [78]. The recovery of TiO_2 and lead derivatives from the perovskite layer is necessary for an environmental benefit but it is not economically attracting for their relatively small contribution on the cost; specifically, TiO_2 can be substituted with SnO_2 and ZnO; Gong et al. [68] reported that the use of ZnO decreases almost every impact. These studies are performed on data obtained for cells and considered scalable to a module.

On the other hand, the recycling of the TCO/glass is convenient for both an economic and an environmental perspective [79]. Another important aspect to consider is that even if the lead emissions during the use-phase are unlikely (if a proper encapsulating strategy is adopted), the occupational exposure and the chronic exposure to lead must be considered [80].

In light of an industrial development of the modules, electrodes made of noble metals will be not feasible [81]: carbon-based electrodes could be the best available solution. Furthermore, the slot-die coating will be preferred to the spin-coating as a perovskite deposition method as the slot-die coating allows material savings. The deposition methods are not analysed in this chapter because the attention has been focused on the materials to comply the extension of this chapter. In a recent study about PSC specifically tailored for a commercial purpose, Celik et al. [81] pointed out the fact the most commonly used organic HTM are very expensive and they do not exhibit a good stability. They suggested a copper-based semiconductor CuSCN as inorganic HTM suitable in an industrial outlook.

The results obtained through many LCA analyses [66, 81] also suggest the importance of increasing the energy efficiency, (especially during the thermal deposition methods), of the PSCs manufacturing process as well the need of decrease the quantity of solvents used.

Substitution of lead with less toxic materials is one of the main challenging purposes for different reasons: (i) the elimination of a cause of concern due to the toxic metal, (ii) the facilitation of the market entry of this new technology also due to the consumer scepticism towards devices containing lead and (iii) the removal of administrative barriers built by the European Union (EU) regarding hazardous materials.

In this way, tin has been suggested as a solution. Nevertheless, it does not seem to be a promising solution. It exhibits ecotoxicity and global warming potential factors higher than lead impacts [53] and, additionally, it is a metal with a low substitution potential. Additionally, its low distribution in the world (Peru, Indonesia, China), which could lead to supply disruption, is dramatically remarkable. Tin is also more expensive than lead, so it could result in a less sustainable PV technology from an economic point of view. Moreover, this evidence is coupled with sensible lower photoconversion efficiency assured by tin-based PSC [64].

Lead is considered an issue for its intrinsic toxicity, but many studies have demonstrated that lead present in PSCs is not the main contributor both for lead emission potential and toxicity potential. Once that the risk and the intensity (lead emission potential) of lead are proved to be admissible, R&D should invest in improve stability, energy efficiency in manufacturing and waste management strategies.

3. Conclusion and perspectives

Organic photovoltaics (OPVs), dye-sensitised solar cells (DSSCs) and perovskites solar cells (PSCs) show promising results regarding costs and environmental performances compared to silicon-based PVs. OPVs and DSSCs still present low efficiencies while PSCs show efficiency values comparable to silicon-based devices.

All the technologies described in this chapter exhibit stability issues and short life-times—the resolution of these drawbacks can only further improve the environmental performances. Pursuing the target of more sustainable emerging photovoltaics, we focused our attention on those that are identified as the hotspots of the studied LCAs. These are: (i) the coated glass, (ii) the precious metals used in the cathode and (iii) the significant use of solvents.

Suggestions to achieve a reduction of the impacts are (i) the use of alternative substrates (e.g. plastic substrates), (ii) the implementation of electrodes not based on precious metals and (iii) the application of the green chemistry principles. Considering the substitution of the coated glass, plastic substrates have been tested and they generally show lower impacts. Yet, is the use of plastic a good idea for the substitution of glass? This question must be asked considering the lower mechanical properties of plastics with respect to glass and the problem of plastic pollution. Regarding the electrodes, carbon-based and inorganic electrodes have been tested and they can be a suitable solution for the industrial fabrication of these emerging technologies.

Solvents are often the most problematic factor in the evaluation of how a process is green. They are used both as the reaction medium and for the purification steps; the huge amount of solvents sensibly affects the green metrics as they produce significant amount of waste. The use of solvents and, in general, the synthesis of new materials, can be regulated through the 12 principles of Green Chemistry [82], with particular attention on the 5th principle as a guideline for solvents.

The use of auxiliary substances (e.g. solvents, separation agents, etc.) should be made unnecessary wherever possible and innocuous when used.

When the minimization of solvents amount is not feasible, the choice of greener ones is the most sustainable practice. Nowadays, charts to choose the greenest solvent are available as those suggested by Prat et al. [83] and Byrne et al. [84]. Also pharmaceutical companies published papers and tables for their selection showing the importance of this topic in the industrial application too [85, 86].

The research cannot pursue only an efficient technology but also a sustainable one, and this is even more important when dealing with alternative energy sources; the energy transition should not be only towards renewable but renewable and sustainable energy. This goal can be achieved only through a responsible research, a production regulated by eco-norms and a properly and thoughtfully designed waste management.

Acknowledgements

The authors acknowledge Mr M. Costamagna for profitable discussion and IMPRESSIVE H2020 project (research and innovation program under grant agreement No 826013) for the financial support.

Conflict of interest

The authors declare no conflict of interest.

Abbreviations

a-Si	amorphous silicon
BOS	balance of system
CED	cumulative energy demand
CPBT	carbon payback time
CRM	critical raw material
DSSC	dye sensitised solar cell
EPBT	energy payback time
ETL	electron transport layer
ETP	ecotoxicity potential
FTO	fluorine doped tin oxide
HTM	hole transport material
HTP	human toxicity potential
ICBA	indene-C60 bis-adduct
ITO	indium tin oxide
LCA	life cycle assessment
m-Si	monocrystalline silicon
OPV	organic photovoltaic
P3HT	poly-3-hexyl-tiophene

PCBM	[6,6]-phenyl-C61-butyric acid methyl ester
PEDOT:PSS	poly(3,4-ethylenedioxythiophene) polystyrene sulfonate
PET	polyethylene terephtalate
PSC	perovskite solar cell
PV	photovoltaic
ROS	reacting oxygen species
TCO	transparent conducting film

Author details

Nicole Mariotti[1], Matteo Bonomo[1,2] and Claudia Barolo[1,2,3]*

1 Department of Chemistry, University of Turin, Turin, Italy

2 NIS Interdepartmental Centre and INSTM Reference Centre, University of Turin, Turin, Italy

3 ICxT Interdepartmental Centre, University of Turin, Turin, Italy

*Address all correspondence to: claudia.barolo@unito.it

References

[1] Vogel JH. The population explosion. In: Paul R. Ehrlich, Anne H. Ehrlich, editors. NewYork: Simon and Schuster; 1990. p. 320. ISBN: 0-671-68984-3. Prometheus 1991; 9:396-397

[2] Bradshaw CJA, Brook BW. Human population reduction is not a quick fix for environmental problems. Proceedings of the National Academy of Sciences of the United States of America. 2014;**111**:16610-16615

[3] Spears D. Smaller human population in 2100 could importantly reduce the risk of climate catastrophe. Proceedings of the National Academy of Sciences of the United States of America. 2015;**112**:E2270-E2270

[4] History. Oil Industry. 2018. Available from: https://www.history.com/topics/industrial-revolution/oil-industry [Accessed: April 29, 2019]

[5] Jaffe RL, Taylor W. The Physics of Energy. Cambridge, United Kingdom: Cambridge University Press; 2018. Available from: https://books.google.it/books?id=sqhJDwAAQBAJ

[6] European Commission. Critical Raw Materials. 2019. Available from: https://ec.europa.eu/ [Accessed:April 10, 2019]

[7] Rebitzer G. Integrating life cycle costing and life cycle assessment for managing costs and environmental impacts in supply chains. In: Seuring S, Goldbach M, editors. Cost Management in Supply Chains. Heidelberg: Physica-Verlag HD; 2002. pp. 127-146

[8] Li B, Gao X, Li J, et al. Life cycle environmental impact of high-capacity lithium ion battery with silicon nanowires anode for electric vehicles. Environmental Science and Technology. 2014;**48**:3047-3055

[9] Parisi ML, Maranghi S, Basosi R. The evolution of the dye sensitized solar cells from Grätzel prototype to up-scaled solar applications: A life cycle assessment approach. Renewable and Sustainable Energy Reviews. 2014;**39**:124-138

[10] PVinsights. Solar PV Wafer Weekly Spot Price. 2019. Available from: http://pvinsights.com/ [Accessed: May 29, 2019]

[11] Tsang MP, Sonnemann GW, Bassani DM. Life-cycle assessment of cradle-to-grave opportunities and environmental impacts of organic photovoltaic solar panels compared to conventional technologies. Solar Energy Materials and Solar Cells. 2016;**156**:37-48

[12] Anctil A, Babbitt C, Landi B, et al. Life-cycle assessment of organic solar cell technologies. In: 2010 35th IEEE Photovoltaic Specialists Conference. 2010. pp. 742-747. DOI: 10.1109/PVSC.2010.5617085

[13] ECHA. European Chemical Agency. 2019. Available form: https://echa.europa.eu/it/home [Accessed: May 15, 2019]

[14] Yannick-Serge Z, Schäffer A, Hugi C, Lenz M, et al. Organic photovoltaics: Potential fate and effects in the environment. Environment International. 2012;**49**:128-140

[15] Ahmad J, Bazaka K, Anderson LJ, et al. Materials and methods for encapsulation of OPV: A review. Renewable and Sustainable Energy Reviews. 2013;**27**:104-117

[16] Brezová V, Gabčová S, Dvoranová D, et al. Reactive oxygen species produced upon photoexcitation of sunscreens containing titanium dioxide (an EPR study). Journal of Photochemistry and Photobiology B: Biology. 2005;**79**:121-134

[17] Anctil A, Babbit CW, Raffaelle RP, Landi BJ. Cumulative energy demand for small molecule and polymer photovoltaics. Progress in Photovoltaics: Research and Applications. 2013;**21**:1541-1554

[18] Espinosa N, Zimmermann YS, Dos Reis Benatto GA, et al. Outdoor fate and environmental impact of polymer solar cells through leaching and emission to rainwater and soil. Energy and Environmental Science. 2016;**9**:1674-1680

[19] Tong Z, Bischoff M, Nies L, et al. Impact of fullerene (C60) on a soil microbial community. Environmental Science and Technology. 2007;**41**:2985-2991

[20] Lyon DY, Alvarez PJJ. Fullerene water suspension (nC60) exerts antibacterial effects via ROS-independent protein oxidation. Environmental Science and Technology. 2008;**42**:8127-8132

[21] Roes AL, Alsema EA, Blok K, et al. Ex-ante environmental and economic evaluation of polymer photovoltaics. Progress in Photovoltaics: Research and Applications. 2009;**17**:372-393

[22] Tsang MP, Sonnemann GW, Bassani DM. A comparative human health, ecotoxicity, and product environmental assessment on the production of organic and silicon solar cells. Progress in Photovoltaics: Research and Applications. 2016;**24**:645-655

[23] Lizin S, Van Passel S, De Schepper E, et al. Life cycle analyses of organic photovoltaics: A review. Energy and Environmental Science. 2013;**6**:3136-3149

[24] Anctil A, Babbitt CW, Raffaelle RP, et al. Material and energy intensity of fullerene production. Environmental Science and Technology. 2011;**45**:2353-2359

[25] Arvidsson R, Svanström M. A framework for energy use indicators and their reporting in life cycle assessment. Integrated Environmental Assessment and Management. 2016;**12**:429-436

[26] Oh WK, Kim S, Yoon H, et al. Shape-dependent cytotoxicity and proinflammatory response of poly(3,4-ethylenedioxythiophene) nanomaterials. Small. 2010;**6**:872-879

[27] Espinosa N, Laurent A, Krebs FC. Ecodesign of organic photovoltaic modules from Danish and Chinese perspectives. Energy and Environmental Science. 2015;**8**:2537-2550

[28] Krebs FC, Dos Reis Benatto GA, Laurent A, et al. Which electrode materials to select for more environmentally friendly organic photovoltaics? Advanced Engineering Materials. 2015;**18**:490-495

[29] Ellen MacArthur Foundation. 2019. Available from: https://www.ellenmacarthurfoundation.org/ [Accessed: May 08, 2019]

[30] Cavallo C, Di Pascasio F, Latini A, et al. Nanostructured semiconductor materials for dye-sensitized solar cells. Journal of Nanomaterials. 2017;**2017**:1-31

[31] Gong J, Liang J, Sumathy K. Review on dye-sensitized solar cells (DSSCs): Fundamental concepts and novel materials. Renewable and Sustainable Energy Reviews. 2012;**16**:5848-5860

[32] Daeneke T, Uemura Y, Duffy NW, et al. Aqueous dye-sensitized solar cell electrolytes based on the ferricyanide-ferrocyanide redox couple. Advanced Materials. 2012;**24**:1222-1225

[33] Yella A, Lee HW, Tsao HN, et al. Porphyrin-sensitized solar cells with cobalt (II/III)–based redox electrolyte exceed 12 percent efficiency. Science. 2011;**334**:629-634

[34] Oskam G, Bergeron BV, Meyer GJ, et al. Pseudohalogens for dye-sensitized TiO_2 photoelectrochemical cells. The Journal of Physical Chemistry. B. 2001;**105**:6867-6873

[35] Financial Times. LME's Cobalt Inquiry Highlights Ethical Issues for Industry. 2017. Available from: https://www.ft.com/ [Accessed: April 03, 2019]

[36] Bella F, Gerbaldi C, Barolo C, et al. Aqueous dye-sensitized solar cells. Chemical Society Reviews. 2015;**44**:3431-3473

[37] Bella F, Ozzello ED, Bianco S, et al. Photo-polymerization of acrylic/methacrylic gel–polymer electrolyte membranes for dye-sensitized solar cells. Chemical Engineering Journal. 2013;**225**:873-879

[38] Bonomo M, Dini D. Nanostructured p-Type Semiconductor Electrodes and Photoelectrochemistry of Their Reduction Processes Energies. 2016;**9**(5):373. DOI: 10.3390/en9050373

[39] Kawano R, Matsui H, Matsuyama C, et al. High performance dye-sensitized solar cells using ionic liquids as their electrolytes. Journal of Photochemistry and Photobiology A: Chemistry. 2004;**164**:87-92

[40] Capello C, Fischer U, Hungerbühler K. What is a green solvent? A comprehensive framework for the environmental assessment of solvents. Green Chemistry. 2007;**9**:927-934

[41] Zhu S, Chen R, Wu Y, et al. A mini-review on greenness of ionic liquids. Chemical and Biochemical Engineering Quarterly. 2009;**23**:207-211

[42] Gontrani L, Bonomo M, Plechkova NV, et al. X-ray structure and ionic conductivity studies of anhydrous and hydrated choline chloride and oxalic acid deep eutectic solvents. Physical Chemistry Chemical Physics. 2018;**20**:30120-30124

[43] Abbotto A, Liliana Boldrini C, Manfredi N, et al. Dye-sensitized solar cells using an aqueous choline chloride-based deep eutectic solvent as an effective electrolyte solution. Energy Technology. 2017;**5**(2):345-353. DOI: 10.1002/ente.201600420

[44] Huang SY, Schlichthörl G, Nozik AJ, et al. Charge recombination in dye-sensitized nanocrystalline TiO_2 solar cells. The Journal of Physical Chemistry. B. 1997;**101**:2576-2582

[45] Lindström H, Södergren S, Solbrand A, et al. Li+ ion insertion in TiO_2 (Anatase). 2. Voltammetry on nanoporous films. The Journal of Physical Chemistry. B. 1997;**101**:7717-7722

[46] Nazeeruddin MK, Kay A, Rodicio I, et al. Conversion of light to electricity by cis-X2bis(2,2′-bipyridyl-4,4′-dicarboxylate)ruthenium(II) charge-transfer sensitizers (X = Cl-, Br-, I-, CN-, and SCN-) on nanocrystalline titanium dioxide electrodes. Journal of the American Chemical Society. 1993;**115**:6382-6390

[47] Law C, Pathirana SC, Li X, et al. Water-based electrolytes for dye-sensitized solar cells. Advanced Materials. 2010;**22**:4505-4509

[48] Lai WH, Su YH, Teoh LG, et al. Commercial and natural dyes as photosensitizers for a water-based dye-sensitized solar cell loaded with gold nanoparticles. Journal of Photochemistry and Photobiology A: Chemistry. 2008;**195**:307-313

[49] Gleick PH, Pacific Institute for Studies in Development, Environment and Security. Water in Crisis: A Guide to the World's Fresh Water Resources. New York: Oxford University Press; 1993

[50] Gleick PH. Global freshwater resources: Soft-path solutions for the 21st century. Science. 2003;**302**:1524-1528

[51] USGS. Water Resources. 2019. Available from: https://www.usgs.gov/ [Accessed: March 06, 2019]

[52] Hui Z, Xiong Y, Heng L, et al. Explanation of effect of added water on dye-sensitized nanocrystalline TiO_2 solar cell: Correlation between performance and carrier relaxation kinetics. Chinese Physics Letters. 2007;**24**:3272-3275

[53] Jung YS, Yoo B, Lim MK, et al. Effect of triton X-100 in water-added electrolytes on the performance of dye-sensitized solar cells. Electrochimica Acta. 2009;**54**:6286-6291

[54] De Wild-Scholten MJ, Veltkamp Petten (Netherlands)] AC [ECN SE. Environmental Life Cycle Analysis of Dye Sensitized Solar Devices. Status and Outlook. Netherlands; 2007

[55] Veltkamp AC, De Wild-Scholten MJ. Dye sensitized solar cells for large-scale photovoltaics; the determination of environmental performances. Renewable Energy. Japan: Makuhari Messe; 2006:9-13

[56] Kakiage K, Aoyama Y, Yano T, et al. Highly-efficient dye-sensitized solar cells with collaborative sensitization by silyl-anchor and carboxy-anchor dyes. Chemical Communications. 2015;**51**:15894-15897

[57] GCell. 2019. Available from: https://gcell.com/ [Accessed: May 22, 2019]

[58] El Baraka A, Baitoul M, Khaldoun A, et al. Development and integration of innovative low-cost PV windows based on dye sensitized solar cells technology: Application in Morocco. In: Proceedings of the 2014 International Renewable and Sustainable Energy Conference IRSEC 2014. 2014. pp. 782-787. DOI: 10.1109/IRSEC.2014.7059866

[59] Ammar AM, Mohamed HSH, Yousef MMK, et al. Dye-sensitized solar cells (DSSCs) based on extracted natural dyes. Journal of Nanomaterials. 2019;**2019**:1-10

[60] Parisi M, Maranghi S, Sinicropi A, et al. Development of dye sensitized solar cells: A life cycle perspective for the environmental and market potential assessment of a renewable energy technology. International Journal of Heat and Technology. 2015;**31**:143-148

[61] Nanowerk. 2019. Available from: https://www.nanowerk.com/ [Accessed: May 22, 2019]

[62] MIT. Energy Initiative. 2011. Available from: http://energy.mit.edu/ [Accessed: May 22, 2019]

[63] Chrkn M. Crystal structure and photoconductivity of cæsium plumbohalides. Nature. 1958;**182**:1436

[64] Grätzel M. The light and shade of perovskite solar cells. Nature Materials. 2014;**13**:838-842

[65] Kojima A, Teshima K, Shirai Y, et al. Organometal halide perovskites as visible-light sensitizers for photovoltaic cells. Journal of the American Chemical Society. 2009;**131**:6050-6051

[66] Zhang J, Gao X, Deng Y, et al. Cradle-to-grave life cycle assessment of solid-state perovskite solar cells. 2017. V004T05A021

[67] Zhang J, Gao X, Deng Y, et al. Life cycle assessment of titania perovskite solar cell technology for sustainable design and manufacturing. ChemSusChem. 2015;**8**:3882-3891

[68] Gong J, Darling SB, You F. Perovskite photovoltaics: Life-cycle assessment of energy and environmental

impacts. Energy and Environmental Science. 2015;**8**:1953-1968

[69] Lee J, Malekshahi Byranvand M, Kang G, et al. Green-solvent-processable, dopant-free hole-transporting materials for robust and efficient perovskite solar cells. Journal of the American Chemical Society. 2017;**139**:12175-12181

[70] Bu T, Wu L, Liu X, et al. Synergic interface optimization with green solvent engineering in mixed perovskite solar cells. Advanced Energy Materials. 2017;7:1700576

[71] Flora G, Gupta D, Tiwari A. Toxicity of lead: A review with recent updates. Interdisciplinary Toxicology. 2012;**5**:47-58

[72] Wani AL, Ara A, Usmani JA. Lead toxicity: A review. Interdisciplinary Toxicology. 2015;**8**:55-64

[73] EC. Directive 2011/65/EU of the European Parliament and of the Council of 8 June 2011 on the restriction of the use of certain hazardous substances in electrical and electronic equipment (RoHS). Official Journal of the European Union 2011;**54**:88-110

[74] Alberola-Borràs JA, Vidal R, Juárez-Pérez EJ, et al. Relative impacts of methylammonium lead triiodide perovskite solar cells based on life cycle assessment. Solar Energy Materials and Solar Cells. 2018;**179**:169-177

[75] Billen P, Leccisi E, Dastidar S, et al. Comparative evaluation of lead emissions and toxicity potential in the life cycle of lead halide perovskite photovoltaics. Energy. 2019;**166**:1089-1096

[76] Kim BJ, Kim DH, Kwon SL, et al. Selective dissolution of halide perovskites as a step towards recycling solar cells. Nature Communications. 2016;7:11735

[77] Huang L, Hu Z, Xu J, et al. Efficient electron-transport layer-free planar perovskite solar cells via recycling the FTO/glass substrates from degraded devices. Solar Energy Materials and Solar Cells. 2016;**152**:118-124

[78] Kadro JM, Hagfeldt A. The end-of-life of perovskite PV. Joule. 2017;**1**:29-46

[79] Binek A, Petrus ML, Huber N, et al. Recycling perovskite solar cells to avoid lead waste. ACS Applied Materials and Interfaces. 2016;**8**:12881-12886

[80] OSHA. Occupational Safety and Health Administration. 2019. [Accessed: May 27, 2019]

[81] Celik I, Song Z, Cimaroli AJ, et al. Life cycle assessment (LCA) of perovskite PV cells projected from lab to fab. Solar Energy Materials and Solar Cells. 2016;**156**:157-169

[82] Grisorio R, De Marco L, Baldisserri C, et al. Sustainability of organic dye-sensitized solar cells: The role of chemical synthesis. ACS Sustainable Chemistry and Engineering. 2015;**3**:770-777

[83] Prat D, Hayler J, Wells A. A survey of solvent selection guides. Green Chemistry. 2014;**16**:4546-4551

[84] Byrne FP, Jin S, Paggiola G, et al. Tools and techniques for solvent selection: Green solvent selection guides. Sustainable Chemical Processes. 2016;**4**:7

[85] Alfonsi K, Colberg J, Dunn PJ, et al. Green chemistry tools to influence a medicinal chemistry and research chemistry based organisation. Green Chemistry. 2008;**10**:31-36

[86] Prat D, Pardigon O, Flemming HW, et al. Sanofi's solvent selection guide: A step toward more sustainable processes. Organic Process Research and Development. 2013;**17**:1517-1525